MW00974784

Modeling and Simulation of ARM Processor Architecture

Mitesh Limachia
Nikhil Kothari

Modeling and Simulation of ARM Processor Architecture

Using SystemC

LAP LAMBERT Academic Publishing

Impressum/Imprint (nur für Deutschland/only for Germany)
Bibliografische Information der Deutschen Nationalbibliothek: Die Deutsche Nationalbibliothek verzeichnet diese Publikation in der Deutschen Nationalbibliografie; detaillierte bibliografische Daten sind im Internet über http://dnb.d-nb.de abrufbar.
Alle in diesem Buch genannten Marken und Produktnamen unterliegen warenzeichen-, marken- oder patentrechtlichem Schutz bzw. sind Warenzeichen oder eingetragene Warenzeichen der jeweiligen Inhaber. Die Wiedergabe von Marken, Produktnamen, Gebrauchsnamen, Handelsnamen, Warenbezeichnungen u.s.w. in diesem Werk berechtigt auch ohne besondere Kennzeichnung nicht zu der Annahme, dass solche Namen im Sinne der Warenzeichen- und Markenschutzgesetzgebung als frei zu betrachten wären und daher von jedermann benutzt werden dürften.

Coverbild: www.ingimage.com

Verlag: LAP LAMBERT Academic Publishing GmbH & Co. KG
Heinrich-Böcking-Str. 6-8, 66121 Saarbrücken, Deutschland
Telefon +49 681 3720-310, Telefax +49 681 3720-3109
Email: info@lap-publishing.com

Approved by: Dharmsinh Desai University, Diss., 2010

Herstellung in Deutschland (siehe letzte Seite)
ISBN: 978-3-659-12088-6

Imprint (only for USA, GB)
Bibliographic information published by the Deutsche Nationalbibliothek: The Deutsche Nationalbibliothek lists this publication in the Deutsche Nationalbibliografie; detailed bibliographic data are available in the Internet at http://dnb.d-nb.de.
Any brand names and product names mentioned in this book are subject to trademark, brand or patent protection and are trademarks or registered trademarks of their respective holders. The use of brand names, product names, common names, trade names, product descriptions etc. even without a particular marking in this works is in no way to be construed to mean that such names may be regarded as unrestricted in respect of trademark and brand protection legislation and could thus be used by anyone.

Cover image: www.ingimage.com

Publisher: LAP LAMBERT Academic Publishing GmbH & Co. KG
Heinrich-Böcking-Str. 6-8, 66121 Saarbrücken, Germany
Phone +49 681 3720-310, Fax +49 681 3720-3109
Email: info@lap-publishing.com

Printed in the U.S.A.
Printed in the U.K. by (see last page)
ISBN: 978-3-659-12088-6

ABSTRACT

Due to the increasing demand of high-end embedded applications such as mobile phones where space, power consumption and cost are major constraints, *RISC* architecture is extensively preferred as compared to *CISC* architecture. The *RISC* architecture makes instructions simpler with fixed-length and easily pipelined in order to achieve single clock cycle throughput even at higher frequencies.

The essential requirements of embedded systems like limited memory, battery power and reduced size of chip led to development of *ARM* processors. Further, *Thumb-2* instruction set architecture in *ARM* enhances performance with higher code densities than previously achievable with the desired balance in order to design the products with tight memory requirements. *ARM Cortex – M3* is a *32-bit RISC* processor based on *Thumb-2* core technology primarily designed to target low cost 32-bit micro controller market.

A software model of a processor core created using *HDL* can serve as the master device to verify functionalities of various slave devices i.e memory or peripherals. The hard model available from the manufacture may not be flexible & compatible with the software platform for the verifications. These typical requirements is supported by *SystemC* which has certain in built function to reduced the complexity of coding different stages of a processor architecture.

This dissertation presents *ARM Cortex- M3* processor model generated using *SystemC* as the modeling and verification tool. Various architecture blocks are modeled in accordance with design specification of *Cortex-M3* processor core. Subsequently they are simulated to verify the behavior of the processor core. Functionality of each block is individually verified using various test conditions. The modules are integrated with validation of functionalities by taking care of timing constraints.

TABLE OF CONTENTS

Page No.

LIST OF FIGURES

vi

LIST OF TABLES

Page No.

ABBRIVATIONS

CISC   Complex Instruction Set Computer.

RISC   Reduced Instruction Set Computer.

MIPS   Microprocessor without Interlocked Pipeline.

PowerPC Power Performance Computing.

GPR   General Purpose Register.

SPARC   From Scalable Processor Architecture.

ARM   Advance RISC Machine.

DUT   Design Under Test.

PDA   Personal Digital Assistances.

ISA   Instruction Set Architecture.

AHB   Advance High performance Bus.

FIFO   First In First Out.

ALU   Arithmetic Logic Unit.

PSR   Program Status Register.

RTL   Register Transfer Language.

ASIC   Application Specific Integrated Circuit

CHAPTER 1

LOW COST & POWER SAVING PROCESSOR ARCHITECTURE

1.1 EVOLUTION OF RISC PROCESSOR

In the early days of the computer industry, programming was done in assembly language or machine code, which required powerful and easy to use instructions. *CPU* designers therefore tried to make instructions that would do work as much as possible. Another goal was to provide every possible addressing mode for every instruction, known as orthogonality, to ease compiler implementation. Data processing operations could therefore often have results as well as operands as a part of memory. The attitude at that time was that hardware design was easier than compiler design, so large parts of the complexity went into the hardware (microcode). Another force that encouraged complexity was very limited memories and also they were quite slow compared to today's cache memories. It was therefore advantageous for the density of information held in computer program to be very high, leading to features such as highly encoded, variable length instructions doing both calculation and data loading. Retroactively, this design philosophy was termed as *Complex Instruction Set Computer (CISC)*.

As *CISC* relies more on hardware for instruction functionality, two principle reasons have motivated the trend; a desire to simplify compliers and a desire to improve performance [1], Consequently, the *CISC* instructions are more complicated and so harder to pipeline or less pipeline. As *CISC* instructions are complex, they consist of variable format instructions set and require multiple cycles for execution. *CISC* processor typically allowed values in

1

memory to be used as operand in data processing instructions. It consists of larger registers set but most had different registers for different purpose. There are only few general purpose registers for storing results or saving operands. In *CISC* more hardware is utilized so size of chip is larger, and consequently the power consumption is higher. *CISC* attempts to reduce the semantic gap between high-level languages and the machine instructions implemented in the hardware. This is done by using instructions that can support many of the constructs found in high-level languages. For example, accessing an element in an array requires the computation of a memory offset from the starting address of the array in memory. Size of the offset is dependent on the size of the array elements and the array index. Complex addressing modes make it possible to access the elements of an array using a single operand specified. For example, consider the following high-level language statement:

$$ARRAYX\ (50) = 0$$

Where, *ARRAYX* is an array of integers starting at address *1000*. An example *CISC* instruction might be:

$$clear.\ word\ 1000\ (r2)$$

Where, register *r2* contains the index (50 in this case). Assume each array element is a word that requires four memory locations, and the starting address is *1000*. To compute the operand address, this instruction first multiplies the contents of the register by four and adds *1000* to it. This philosophy attempts to reduce execution time by minimizing the path length or number of instructions executed [2]. In *CISC* architecture, major drawbacks are increased size of chips, high power consumption and multiple cycle execution. To overcome these limitations and strength the performance of core, RISC architecture was evolved as discussion in the following section.

2

1.1.1 RISC Design - Introduction

In *1970s* researchers at *IBM* demonstrated that the most of combinations of orthogonal addressing modes and complex instructions were not used by most programs generated by compilers used at that time. It was also found that, on microcode implementations of certain architectures, complex operations observed to be slower than a sequence of simpler operations performing the same task. One example was the *VAX's INDEX* instruction, which ran slower than the equivalent implementation using simpler operations [3].

It was also observed that instead of focusing on minimizing the path length, if we reduce the average number of cycles per instruction that can be achieved by implementing just the simple instructions in hardware. An added benefit is that simpler hardware is required so a shorter clock period is feasible. However, using only simpler instructions means the path length will increase because complex operation perform the same task in one instruction, so this approach require several instructions to perform the same task, however this effect is small as compared to reduction in average instruction cycles and cycle time[4]. Since most of programs spend most of their time to execute simple operations, so in order to increase execution speed these operations were made such that they were simpler and faster in nature. However, clock rate of CPU is limited by the time it takes to execute the slowest sub-operation of any instruction. Decreasing the clock cycle time often improves the execution of subsequent instructions.

One other important observation made by studying real word program is that most of all the constants used in program would fit in 12 to 13 bits, but many CPUs used 16 or 32 bits to store them, which means processors often

3

had oversized immediate. This observation suggests that in order to reduce the number of memory accesses, fixed length op-codes were defined that store constants/immediate in unused bits of instruction word so when processor requires, it would be immediately ready and would increase speed of execution. However, this required small op-codes in order to provide room to store sized immediate/constants in 32-bit instruction word. The focus on *reduced instructions* led to the resulting machine being called a *Reduced Instruction Set Computer.* Here, the goal was to make instructions simpler, fixed- length and easily pipelined in order to achieve a single clock cycle throughput still working at higher frequencies. Later, it was found that one of the most significant characteristics of *RISC* processors was that access to external memory would only be possible by either load or store instructions, remaining all other instructions of Instruction Set Architectures (ISA) were limited to access only internal registers. This feature simplify processor design in many aspects, allowing all instructions to be fixed length, simplifying the pipelines, most of all operations restricted to internal registers, delay in memory access dealing with only two instructions. This led to RISC design being referred to as load/store architecture [5].

A common misunderstanding of the term *reduced instruction set computer* is the mistaken idea that instructions are eliminated so resulting in smaller set of Instruction Set Architecture (ISA). In fact, over the years, *RISC* instruction sets have grown in size, and today many of them have a larger set of instructions than many *CISC CPU*s[6][7]. The term *reduced* was used to describe the fact that the amount of work that any single instruction contains is reduced as compared to the *complex instructions* of *CISC CPU* that required number of data accesses memory cycle in order to execute a single instruction [5]. *RISC* architecture consists of fewer and simple instructions. It relies more on software compared to hardware and

4

consequently the *RISC* instructions are simple and easy to pipeline. As *RISC* instructions are simple, it consists of fixed format instruction sets and requires only single cycle for execution. *RISC* processors typically allowed values in registers to be used as part of operand so speed of execution is faster. In RISC, most of all the registers are general purpose registers except program counter. As lesser hardware is utilized, size of chip is smaller and the power consumption is reduced.

RISC design philosophy is based on four major design rules:

- Instructions: *RISC* processor consists of reduced number of instruction categories that provides simple instructions, which can execute in single cycle. Here, each instruction has fixed length that allows the pipeline to fetch next instruction while previous instruction is in next (decode) stage.

- Pipeline: It is a parallel mechanism that ideally advances by one step in each cycle for maximum throughput.

- Registers: It contains larger general purpose register set. Any register can contain either data or an address. Register is used to store content of operands and it acts as a faster local memory storage device.

- Load-Store architecture: Processor operates directly on data held in registers. Separate Load and Store instructions available to transfer data between the register bank and memory.

Followings are the key features of *RISC* architectures [8]:

- A single cycle execution.

- A fixed instruction set (*32-bit*) with only few categories.

- A load-store architecture, where instruction processes directly on registers and separated from instructions that can accesses the main memory.

- Pipelined executions.

- Hard-wired instructions decode logic.

- Delayed branch slot.

1.1.2 CISC versus RISC

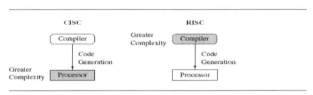

Fig. 1.1 CISC vs. RISC

As shown in Fig. 1.1, *CISC* emphasizes more on hardware complexity while RISC emphasizes more on complier complexity [9].

Table 1.1 shows the comparison between *CISC* and *RISC* architectures:

Table 1.1 Comparison between CISC and RISC architectures

Performance Indicator	CISC	RISC
Instructions	Complex and different	Simpler and similar
Number of cycles of execution	Multiple	Single
Pipelined	Less or Harder	Highly
Format of Instructions set architecture	Variable	Fixed
Complexity	Micro- program	Complier
Size of chip & Power consumption	Large & High	Small & Low
General Purpose Registers	Few	Most of all

6

1.1.3 Processor Performance Indicator

Followings are the three factors, which point out why *RISC* processors are able to outperform *CISC* processors.

- Path Length: *RISC* processors need more number of instructions than *CISC* processors to perform a same task. Studies show that this is about a factor of *1.2* more instructions [2].

- Cycles/ Instructions: The goal of *RISC* processor is to execute one instruction per cycle but because of cache misses, not filling all the branch delay slot, the average is slightly greater than one cycle/instruction and close to *1.3* [2].

- Clock periods: Although it is technology dependent but by looking at RISC processors currently available in market, we can estimate that RISC machine could run about *1.5* times faster than *CISC* machine [2].

By considering above parameters, in order to compute execution time using following equation-(2.1)

Execution Time = Path Length (Instructions) × Cycles/ instructions × Clock Period (seconds/cycle)   (2.1)

It was found that execution time of *RISC* processor is about *3.8* times that of CISC processor. This suggests that a *RISC* processor has performance advantage of about factor of four as compared to *CISC* processors approximately [2].

RISC merits:

- A small die area: Simple processor would require fewer transistors and less silicon area.

7

- A shorter development time: A simple processor should take less design efforts and therefore having a low design cost.

- A higher performance.

RISC demerits:

- More instructions: Requires great demand in instruction memory and secondary storage.

- Complex software: To handle implementation details such as pipeline hazards, design of complier must be optimized and efficient.

- *RISC* generally has poor code density as compared to *CISC*.

- *RISC* don't execute x86 code.

- High memory bandwidth: It is a measure of how fast data can be transferred between memory and CPU.

1.2   RISC- BASED PROCESSOR ARCHITECTURES

RISC architecture was adopted by number of processor families like *MIPS, PowerPC, SPARC*, and *ARM* etc. [10] [11] [12]. All of them adopted the basic features from *RISC* architecture and added their own features to improve performance of the core. In *MIPS (Microprocessor without Interlocked Pipeline)*, a major aspect is to fit every sub-phase of all instructions in to one cycle, thereby removing any need for interlocking and permitting a single cycle throughput. The other difference was handling of subroutine calls to eliminate interlocking. For this purpose, more number of registers were used, which not only made this task simple but also increase the performance of all tasks [10]. In *PowerPC (Power Performance*

8

Computing), the ISA is divided into several categories and every component is defined as a part of category. Each category resides within a certain Book. Processors implement a set of these categories. It is *RISC* load/store architecture with multiple set of registers i.e. thirty-two *32-bit* or *64-bit* General Purpose Registers (GPRs) for integer operations etc. [11]. In *SPARC(from Scalable Processor Architecture)*, *a*im is to execute instructions at a rate of almost one instructions per clock cycle with lack of instructions such as multiply or divide, which made them similar to *MIPS* architecture. *SPARC* processor usually contains as many as 128 general purpose registers Among them only 32 are visible to software. The *Scalable* in *SPARC* comes from the fact that the SPARC specification allows implementation to scale from embedded processors up through large server processors, all sharing the same core instruction set. Only parameters that can be scaled are number of implemented register windows [12]. In *ARM (Advanced RISC Machine)* architecture to make design simple and fast, best features from *RISC* architecture were adopted. In addition *ARM* has included features like conditional execution of most instructions, use of 32-bit barrel shifter, a link register for fast leaf function calls, powerful indexed addressing modes. *ARM* processor was evolved in order to meet requirements of embedded systems like small size, higher performance and low cost [9].

1.3 ARM - MODIFIED RISC ARCHITECTURE

Advanced *RISC* Machine or *ARM* processor is designed based on *RISC* principle. *ARM* core is not a pure *RISC* architecture because of constraint of its primary applications – the embedded systems. The *ARM* processor has been specifically designed to be small to reduce power consumption and extend battery operation, which is essential for applications such as mobile phones, modems and personal digital assistants (PDA's). High code density is another major criterion since embedded systems require limited memory due to cost and/or physical size restrictions, high code density is useful for applications that have limited on-board memory, such as mobile phones and mass storage devices. For high-volume applications like mobile phones, every cent has to be accounted for in the design. The ability to use low-cost memory devices produces substantial savings. Another important requirement is to reduce the area of the die taken up by the embedded processor. For a single-chip solution, the smaller the area used by the embedded processor, the more available space for specialized peripherals. This in turn reduces the cost of the design and manufacturing since fewer discrete chips are required for the end product. *ARM* has incorporated hardware debug technology within the processor so that software engineers can view what is happening while the processor is executing code. With greater visibility, software engineers can resolve issues faster, which have a direct effect on the time to market and reduce overall development costs. *ARM7TDMI, ARM11* and all Cortex series processors are the most extensively used ARM processors.

ARM architecture incorporated a number of features from *RISC* design listed as follows:

- Load-Store architecture: To reduce memory access.

- Fixed -length *32 bit* instructions: In order to make instructions simpler.

- 3- address instructions format

- Uniform *16 x 32 bit* register file: For quickly accessing the operands.

- Mostly single cycle execution: To increase speed of operation.

ARM had avoided number of features from *RISC* design listed as below:

- Register windows: On cost background.

- Delayed branches: Efficient pipelining.

- Single cycle execution for all instructions: To reduce memory access.

Additionally, ARM has introduced certain features like:

- Variable cycle execution for certain instructions: – To transfer multiple data from memory using a single instruction.

- Inline barrel shifter leading to more complex instructions- To improve computation power.

- Thumb 16-bit instruction set: This is a revised instruction set and a subset of ARM instruction set encoded in to 16 bits in which each Thumb instruction having exact equivalent 32-bit *ARM* instruction requiring dynamic decompression hardware in the instruction pipeline.

- Conditional execution of an instruction: For efficient pipelining.

- Link register for fast leaf function calls: To optimize subroutine and exception processing.

- Powerful indexed addressing modes.

ARM processors also have some features rarely seen in inter *RISC* architecture, such as PC-relative addressing and pre and post-increment addressing modes. With a 32 bit wide memory access, 16-bit Thumb code gives a lower performance than *ARM* 32 bit code because more 16 -bit instructions must be executed to perform the same task [13].

1.4 ARM CORTEX– M3 PROCESSOR

To improve code density and performance of core, Thumb-2 instruction set was introduced in which memory footprint usage is just *74%* of an *ARM-* equivalent implementation [15]. It also has the benefit of eliminating the switching between *ARM* and Thumb state to speed up execution. Thumb-2 core technology is a significant enhancement to the *ARM* architecture, which provides performance at higher code densities than previously achievable with *ARM* architecture. Due to that all *ARM* Cortex profile micro controller i.e. *ARM Cortex- A series, ARM Cortex – R series, ARM Cortex - M* series had adopted Thumb-2 core technology [14]. M-profile processors are especially developed for targeting low cost application in which low gate count, processing efficiency, low interrupt latency and ease of use are critical. *ARM Cortex- M3* is a M-profile processor based on *ARM v7-M* architecture ideally suited for applications like Micro Controller Unit, wireless network, automotive and industrial control system, white goods and medical instrumentation [14]. The device blends the best features from the 32-bit *ARM* architecture with the highly successful Thumb-2 instruction set design while adding several new capabilities [14].

ARM Cortex- M3 processor consists of three stage operations. Fetch, decode and execute stages perform in single cycle as per the pipeline behavior adopted from the *RISC* architecture. When a branch instruction is encountered, the decode stage also includes a speculative instruction fetch that could lead to faster execution. The processor fetches the branch destination instruction during the decode stage itself. Later, during the execute stage the branch is resolved and it is known which instruction is to be executed next. If the branch is not to be taken, the next sequential instruction is already available. If the branch is to be taken, the branch instruction is made available at the same time as the decision is made, restricting idle time to just one cycle. The *Cortex- M3* core contains a decoder for traditional Thumb and new Thumb-2 instructions, an advanced *ALU* with support for hardware multiply and divide, control logic, and interfaces to the other components of the processor [16].

Processor consists of 3- word entry pre-fetcher Unit. It pre-fetches at one word/cycle from memory through *I-CODE* Bus. To increase the strength of core, load/store address computation unit is combined with decode stage while write back and load/store data storage operations are combined with execution unit [17]. The processor incorporated Harvard architecture to speed up the execution of instructions by performing simultaneous instruction fetches with data load/store.

For modeling of *ARM Cortex- M3* processor, different modeling techniques were reviewed as discussed in following section.

1.5 MODELING TECHNIQUES

In High Level Description Language (*HDL*), the different modeling techniques are used. Behavioral modeling allows module be implemented in terms of desired design algorithm without concern for the hardware implementation. In data flow modeling technique, module is specified by data flow. In Gate level modeling, module is implemented in terms of logic gates and interconnection between them. Switch level modeling allows module be implemented in terms of switches or storage nodes [18].

A *HDL* language *SystemC* makes the behavioral modeling easier as discussed in the next section.

1.6 SYSTEMC

SystemC is both a system level and hardware description language. It is a single language for modeling hardware and software systems. *SystemC* is based on *C++* programming language. It extends the capabilities of *C++* to enable hardware description and adds important concepts such as concurrency, events and data types. These capabilities are provided via class library. The class library is not a modification of *C++*, but a library of functions, data types and other syntax that are legal *C++* code. Class library also provides new mechanism to model system architecture with hardware timing, concurrency and reactive behavior. These Classes enable the user to define modules, process and communication through ports and signals that can handle a wide range of data types ranging from bits, bits vectors to standard *C++* types to user- defined data types such as enumeration types and structure types [19]. *SystemC* also provides syntax that describes concepts familiar to hardware designers such as signals, modules and ports. Additionally, *SystemC* offers a simulation kernel that allows us to simulate an executable specification of design or systems. A unique thing that *SystemC* provides is that the same language can be used for followings:

- System level design.
- Describing hardware architectures.
- Describing software algorithms.
- Verification.
- *IP* exchanges.

We can describe overall system by using *SystemC*. *SystemC* consists of

14

certain inbuilt functions, which are extremely useful during modeling of any architecture. It contains rich data types; for an example *sc_lv* data type, which supports *X* values. Furthermore, *SystemC* is an open source and freely available language. *SystemC* is powerful enough to describe all kinds of verification environments from the signal level to transaction level. *SystemC* also allows for test bench generation and testing. It can serve as a verification language as well. In this project, *SystemC 2.2.0* version is used with kernel version of *2.6.18* to simulate the Design Under Test (*DUT*). For compilation, *GCC* complier is used. For viewing the graphical results and debugging purpose, waveform viewer *GTKWave* analyzer with version v *3.2.2* is used.

1.7 MAIN CONTRIBUTION

During the project, various activities were carried out as listed below:

- Survey of *RISC* architectures.

- Survey of *RISC-* based processors architecture.

- Study of *ARM* processors architecture.

- Study of *ARM Cortex- M3* pipeline architecture, features and application.

- Survey of *SystemC* as a modeling language.

- Design of architecture blocks for different individual units/stages by considering various control signals.

- Successfully modeled different blocks of *Cortex- M3* processor like Prefetch/Fetch Unit, Thumb Instruction decoder and control unit, shifter Unit, *ALU* unit, Register file, Advanced High Performance Bus (*AHB*) and Reset exception using *SystemC*.

- Functionality of each different modules/Units was verified by passing different test cases through test bench and viewing results on waveform viewer.

- Integrated the designed modules and again verified the functionality as per the timing requirements. During coding of different modules of processors various design challenges were faced like, the numbers of input and output ports accomplished by register file, utilization of each buffer among 3-word entry Prefetcher Unit, Op-code encoding of Thumb-2 Instruction Set Architecture, destination register number sent to the register file before or during execution of an instruction, clock cycle in which destination and *PSR* register are required to update, pipeline reloading operation in the case of execution of conditional and unconditional branch instruction etc.

1.7.1 Scope of Dissertation

The scope of this dissertation is restricted to generate a model of the processor by excluding co-processor and debugging peripherals. Therefore, the verification has employed the relevant Thumb and Thumb-2 instructions only. Modeling of exceptions (except Reset) is excluded in the project.

1.8 DISSERTATION OUTLINE

Chapter 1 introduces the *RISC* architecture and its benefits in the applications requiring low power and low cost designs. Various processor based on *RISC* architectures are briefly explained in chapter. The chapter also discusses *ARM* family of processors and the modeling techniques for them.

In *Chapter 2*, main focus is on study of evolution of *ARM* processors, *ARM* as a modified *RISC* architecture, *ARM* core data flow model, programmer model, pipelined characteristic and data processing instruction execution. Various *ARM* processor family comparisons are briefly described.

Chapter 3 has the main focus on introduction of *ARM Cortex –M3* processor, its features, applications, 3-stage pipelined architecture, and comparison with famous *ARM7TDMI* processor.

Chapter 4 presents modeling of *ARM Cortex- M3* processor containing different architecture blocks of processor, flow charts of different blocks and pseudo code related to different functionalities.

Chapter 5 containing main focus on simulation environment used for modeling and simulation results of different stages/units by applying various test vectors.

Chapter 6 presents conclusion and future scope for the work.

1.9 SUMMARY

In this chapter, *CISC* and *RISC* design aspects, typical characteristics of RISC architectures with advantages and drawbacks are discussed in brief. Different *RISC* based processor architectures, *ARM* as a modified *RISC* architecture are also discussed. Evolution of *ARM Cortex- M3* processor and its architecture, different modeling techniques, *SystemC* as a widely used modeling language and its features are briefly described. The chapter discusses the major contribution of the project. Finally, outline of the dissertation is provided.

CHAPTER 2

ARM - ADVANCED RISC MACHINE

2.1 EVOLUTION OF ARM PROCESSORS

There are number of features and requirements of embedded design that has driven the *ARM* processor design. Essential requirement of an embedded system is some form of battery power – that is *ARM* processors designed in such a way that they would be small in size to reduce power consumption and provide extended battery operation that is essentially useful in application like mobile phones, digital camera etc. As embedded systems have limited memory due to constraint of cost as well as physical area, high code density is also an important requirement in the embedded system applications. High code density is always critical for those applications that have small memory like *PDA's* and mass storage devices. Another major requirement is to reduce size of chip that in turn reduces the cost of design and manufacturing since fewer chips are required for the end product. The *ARM* is a *32-bit RISC* processor architecture developed by *ARM* Limited that is widely used in embedded designs. The first *ARM* processor was developed at Acorn Computers Limited, of Cambridge, England, between *October 1983* and *April 1985*. At that time, and until the formation of Advanced *RISC* Machines Limited, *ARM* stood for Acorn *RISC* Machine.

To design products with low cost and low power consumption like mobile equipments and mass storage device, different *ARM* family processors have been introduced as shown in Fig. 2.1, which shows the technology evolution of the ARM product family by considering the evolution of Instruction Set Architecture (ISA) and evolution of different core versions.

18

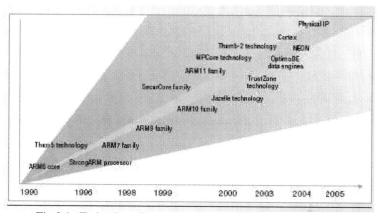

Fig 2.1   Technology Evolution of the ARM Product Family

There are number of *ARM* processor core existing as per industry requirements, among them most famous *ARM* core was *ARM7- TDMI* that was essentially useful in embedded applications, where major concerns are low power consumption and high code density.

2.2   ARM- MODIFIED RISC ARCHITECTURE

ARM core is not a pure *RISC* architecture because of the constraints of its primary application—the embedded system. In some sense, the strength of the *ARM* core is that it does not take the *RISC* concept too far. The *ARM* architecture has adopted number of features from *RISC* architecture design, but a some of the features are dropped. The *ARM* architecture includes the following *RISC* features:

- Load/Store architecture.
- Uniform *16 × 32* register file.
- Fixed instruction width of *32 bits* to ease decoding and pipeline.
- Mostly single cycle execution.

19

- No support for misaligned memory accesses: Code/Data fetched from memory are always word aligned.
- 3-address instruction formats.

The ARM architecture has avoided followings features that are employed on the Berkeley *RISC* designs:

- Delayed branches- It is a way to increase the efficiency of the pipeline, make use of a branch that does not take effect until after execution of the following instruction. The instruction location immediately following the branch is referred to as the delay slot [21], the instruction in this slot is executed whether or not the branch is taken in other words the effect of the branch is delayed. This instruction keeps the *ALU* of a processor busy for extra time than normally needed to perform a branch. The problem with delayed branch is that they remove the atomicity of individual instructions. It works well on single issue pipeline processors, but not with super scalar implementation. Delayed branch also made exception handling more complex. Single cycle execution of all instructions- Single cycle execution of all instructions is only possible if there are separate instructions/codes and data memories, which are too much expensive for *ARM* application which was preliminarily developed for embedded systems. Instead of single cycle execution of all instructions, the *ARM* was designed to use the minimum number of cycles required for memory accesses. When essential, extra cycles and auto-indexing addressing modes are used. These reduces the total number of *ARM* instructions to perform any operation and improves performance and code density of core. *ARM* executes most data movement and data processing instructions in single clock cycle except a few instructions that take multiple cycles for executions.

- Register windows - A normal *CPU* has a small number of registers and program can use any registers at any time. In a *CPU* with register windows, there is a huge number of registers e.g. *128* but executable program can use only a small number of them e.g. 8 at a time. A program, which limits itself to 8 registers per procedure can make very fast procedure calls. The call simply moves the window *down* by 8, to the set of 8 registers used by those procedures, and the return moves the window back. This feature actually reduces the traffic between the processor and memory resulting from register saving and restoring. *ARM* dropped this feature on cost ground as large number of registers would occupy large chip area. Although *ARM* uses shadow registers to handle exception, that is not too much different in concept.

To meet the target requirements of embedded systems and to improve performance of the core, *ARM* has its own additional features as followings:

- Conditional execution of instructions- This feature is used to reduce branch overhead and compensating for the lack of a branch predictor. Here, *4- bit* condition code is used on the front of every instruction, meaning the execution of every instruction is optionally conditional. Other *CPU* architectures typically have condition codes on branch instructions. This cuts down significantly on the encoding bits available for displacements in memory access instructions but on the other hand it avoids branch instructions when generating code for simple if statements. This results in a typical *ARM* program being denser than expected and fewer memory accesses so the pipeline is used more efficiently. This feature improves performance and code density by reducing branch instructions.

21

- Variable cycle execution for certain instructions - Not all *ARM* instructions execute in a single cycle. *load –store* instructions with multiple registers require more than one cycle for execution depending upon the number of registers require to be loaded from memory. Over here, execution cycle depends upon the number of registers being transferred. In case of multiple registers transferred, transfer can occur on sequential memory address because sequential memory access is always faster than random access that increases performance of the core. Generally load-store operation with three registers transfer requires four cycles; one cycle for pipeline execution and remaining depends upon the number of registers being transferred.

- Inline barrel shifter - Barrel shifter is a purely combinational shifter that preprocesses one of the operands, generally operand 2, before it is used by instruction. Barrel shifter is used at no extra cost in terms of code with little additional cost in terms of time. It improves performance of the core by improving code density and has ability to multiply or divide any number with power of 2^n. It is also used for address offset calculation during load-store related operations. This feature improves code density of many instructions.

- Thumb 16-bit instruction set - High-end embedded control applications such as cell phones, disk drives and modems are demanding more performance from their controllers at low cost. By implementing a second *compressed* instruction set termed as a Thumb that reduces *RISC* code size improves *32-bit ARM RISC* performance and power consumption at *8/16bit* system cost [13]. Thumb instruction set was a subset of *ARM* instruction set encoded into its 16 bits as shown in Fig. 2.2. Each Thumb having

22

an exact equivalent *ARM* instruction makes decompressing simple.

Three types of *ARM* instruction were considered in the search for code compression; those which benchmarking customer code has shown to be the most frequently used and hence the most critical to shrink, those needed by the compiler to produce compact code and those that has some redundancy in the fixed length Op-code. The final Thumb instruction set contains a subset of 36 instruction formats drawn from the standard *32-bit ARM* instruction set that have been re-coded into *16-bit* wide op-codes. The net result of compressing code in this way is typically a *30%* improvement in code density over native *ARM* code [13]. Due to 16 bit instruction width limitation, all of the features of the *ARM* instruction set are not present in the Thumb instruction set. As a result, more *16-bit* instructions than 32 bit instructions are required to execute a given function. However, as Thumb instructions are half the width of 32-bit *ARM* instructions, the 16 bit code is on average *35%* smaller [13]. While accessing 32-bit wide memory system, 16-bit Thumb instruction set provides lower performance than *ARM* 32-bit instructions because more 16-bit Thumb instructions must execute to perform the same task.

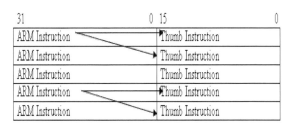

Fig. 2.2 ARM to Thumb Instruction set Mapping.

While using an 8-bit or 16-bit memory system, Thumb code always provides higher performance as compared to 32-bit *ARM* instructions because twice memory access is required to fetch 32-bit instructions from memory than to fetch a single 16-bit Thumb instruction. Since Thumb code are able to execute standard *ARM* instructions as well as the new Thumb instructions, the designer can trade-off code size against performance by writing size-critical routines in Thumb code and performance- critical routine in ARM code.

- A link register for fast leaf function calls - Generally register *R14* is used as a link register in *ARM* processors. It is used to store return address of an instruction during execution of branch instruction or exception. It improves the processing speed by storing the return address.

- Powerful indexed addressing modes - *ARM* processors are using PC-relative addressing as PC is one of its *16* registers, *pre* and *post* increment addressing modes in the case of load-store related instructions.

- Enhanced instructions - The enhanced digital signal processor instructions were added to the standard ARM instruction set to support fast *16×16* multiplier operations and saturation that improves performance of the core.

These additional features made *ARM* processor one of most extensively used embedded processor cores targeting low cost *32-bit* microcontroller market.

2.3 ARM CORE DATAFLOW MODEL

Fig. 2.3 represents *ARM* core data flow model. It also represents the components which are essential to make *ARM* core.

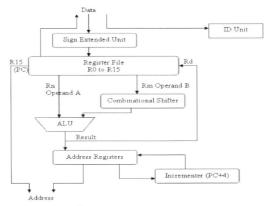

Fig. 2.3 ARM Core Data Flow Model

The main components are instruction decoder, Register file, Barrel shifter, Arithmetic Logic Unit (*ALU*), Multiply- Accumulate Unit (*MAC*), Address register, Incrementer and buses.

As shown in Fig.2.3, an *instruction* or *data* enters through data bus. Here, Von Neumann architecture is used for *ARM* core, where instruction or data share the same bus. First, the instruction belonging to Instruction Set Architecture (*ISA*) is decoded by instruction decoder, which is a purely combinational device generating function code for each of decoded instruction that instructs the execution unit to perform the specified task and also generates the necessary control signals. As *ARM* processor uses a load-store architecture, all instructions process the data directly available in the internal registers except the load and store instructions transferring data in and out from processor and communicating with memory. There are no data processing instructions to directly manipulate the data in memory. So, data

25

processing is carried out in registers. Register file is a bank of *32* registers used to store the data items. Source operands values are read from the register file. *ARM* instructions typically consist of two source operand registers *Rn* and *Rm* and one destination operand *Rd*. Values of source operand registers *Rn* and *Rm* are read from the register file and value of destination register is written into the register file. After reading the source operands from the register file, both the operands are sent to execution stage via different buses. One of the operands register *Rm* is preprocessed in barrel shifter before it enters the *ALU*. Thus, both barrel shifter and *ALU* can compute a wide range of address and calculations. The *ALU* takes the operands *Rn* and *Rm* from *A* and *B* buses, processes them and computes the result. It also writes the result in *Rd* directly to register file using Result bus. Load-Store instruction can also use the *ALU* to generate an address to be held in address register. For Load-Store instructions, incrementer updates the address registers before core reads or writes the next register value from or to the next sequential memory location. The processor continues to do this operation until any branching instruction or an exception changes the normal execution flow.

2.4 LOAD –STORE ARCHITECTURE

ARM architecture has adopted load-store architecture from *RISC* architecture. In ARM ISA, instruction will only process values, which are stored either in registers or specified directly within the instruction itself. It always places the results of such processing into a register again. The only instructions, which transfer data to and from memory state are load and store instructions. Among them, load instructions copy memory values into registers and store instructions copy register values into memory. *CISC* processors typically allow a value from memory to be added to a value in a register and sometimes allow a value in a register to be added to a value in memory. *ARM* does not support such memory to memory operations. Therefore, all *ARM* instructions fall into one of the following three categories:

Data processing instruction: These category of instructions can use and change only register values. For example, an instruction can add two registers and place the result in a register.

Data transfer instructions: These categories of instructions copy memory values into registers (load Instructions) or copy register values into memory (store instructions). An additional form, useful only in systems code, exchanges a memory value with a register value.

Control flow instructions: Normal instruction execution uses instructions stored at consecutive memory addresses. Control flow instructions cause execution to switch to a different address, either permanently (branch instructions) or saving a return address to resume the original sequence (branch and link instructions) or trapping into system code (supervisor calls).

2.5 ARM PROGRAMMER'S MODEL

REGISTERS:

R 0	R 1	R 2	R 3	R 4	R 5	R 6	R 7	R 8	R 9	R 10	R 11	R 12	R 13	R 14	R 15	C P S R	S P S R
													SP	L R	P C		

Fig. 2.4   Registers Used in ARM Core.

27

A processor's instruction set defines the operations that the programmer can use to change the state of the system using the processor. This state usually contains the values of the data items in the processor's visible registers and the system's memory. Each instruction can be viewed as performing a defined transformation from the state before the instruction is executed to the state after it has completed. Function of general purpose register is to hold data or an address. As shown in Fig. 2.4, there are *18* 32-bit wide registers. Among *them, 16* are data and two are program status registers. The data registers are visible to programmer as *R0 to R15*. Among data registers, three registers *R13, R14, R15* are used for a particular task or special function as followings:

Register *R13* is used as a stack pointer (*SP*) to store the top of stacks.

R14 is called Link Register (LR) and used to store program return address during execution of a subroutine or branching instructions.

R15 is the program counter (PC), which contains the address of the next instruction to be fetched by the processor.

R0 to R13 are orthogonal registers i.e. any instruction that can apply to register *R0* can also apply to any other registers, depending upon the context. *R13 and R14* area also useful as a general purpose registers. There are two program status registers: Current Program Status Register (*CPSR*) and Saved Program Status Register (*SPSR*). The register file contains all registers available to a programmer.

Current Program Status Register (*CPSR*):

31			28	27		7	6	5	4		0
N	Z	C	V	Unused			I	F	T	Mode	

Fig. 2.5 Current Program Status Register (CPSR) Format.

The conditional flags are in top four bits of the register and have the

28

following meanings:

N: Negative, this flag is set if top bit of 32-bit result is one. The last *ALU* operations that changed the flags produced a negative result.

Z: Zero, this flag is set if every bit of 32-bit result is zero. The last *ALU* operations which changed the flags produced a zero result.

C: Carry the last ALU operation which changed the flags generated a carry-out, either as a result of an arithmetic operation in the *ALU* or from the shifter.

V: Overflow; the last arithmetic *ALU* operation which changed the flags generated an overflow into the sign bit.

I: I = 1 disables IRQ interrupts.

F: F= 1 disables FIQ interrupts.

T: T= 1 indicates Thumb state. *T* = 0 indicates ARM state.

Mode: Indicates the current processor mode.

2.6 ARM INSTRUCTION SET

All *ARM* instructions are *32-bit* wide except 16-bit wide Thumb instructions are aligned on *4-byte* boundaries in memory. The most remarkable features of *ARM* instructions set are as follows [25]:

- The load-store architecture: Only load-store instructions can access the main memory i.e. *LDR, STR* etc.

- 3–address data processing instructions: All data processing instructions consist of three registers. Among them two are source operand registers and remaining is result or destination register.

- Conditional execution of every instruction: All instructions contain 4-bit conditional code in front that determines execution of an instruction.

- Inclusion of load and store with multiple registers: For an example, *LDMIA/STMIA* with auto indexing mode.

- Ability to perform a general shift operation and a general *ALU* operation in a single instruction that executes in a single clock cycle.

- Very dense *16-bit* compressed representation of the instruction set in the Thumb architecture.

For small embedded systems, which use *ARM* processors, code density advantage outweighs the small performance penalty incurred by the decode complexity. Thumb code extends this advantage to give *ARM* better code density than most *CISC* processors.

2.7 PIPELINE BEHAVIOR AND INSTRUCTION EXECUTION

Fig. 2.6   Three-stage pipeline ARM core.

Pipeline is generally used to speed up the execution by fetching the next instruction while other instructions are being decoded and executed. It is a parallel mechanism used to improve performance of the core by allowing core to execute instruction in every cycle. *ARM* adopted pipeline mechanism from *RISC* architecture. Three stages of pipeline as shown in   Fig. 2.6 as follows:

Fetch stage: This stage is used to fetch instruction from memory through bus

30

architecture. It determines the correct instruction going for execution.

Decode stage: In this stage, instruction is decoded, function code is determined and necessary control signals are generated to perform the subsequent operations on operands.

Execution stage: In this stage, instruction execution takes place as per the function code, address computation is also done if required and result is written_back to register file.

As the length of pipeline increases i.e. five stage pipeline, the amount of work required to be done at each stage is reduced, which allows the processor to work at a higher operating frequency. This in turn increases the performance [8]. The system latency increases because it takes more cycles to fill the pipeline before the core can execute an instruction. The increased pipeline length also means there can be data dependency between certain stages. The major complexity with increased length of pipeline is data forwarding. As instruction execution is spread across multiple stages, to resolve data dependency without pipeline stalling is to define forwarding path. Forwarding path allows results to be forwarded as soon as available.

2.8 ARM PROCESSOR FAMILIES

ARM has designed number or different processors that are grouped into different families as per use of the core. The families are based on the *ARM7, ARM9, ARM10 and ARM11* cores. The ascending number equates to increase in performance and sophistication. One of the most significant changes to *ISA* was introduction of the Thumb set in *ARM7TDMI* processor.

Table 2.1 shows the comparison between various *ARM* families like *ARM7, ARM9, ARM10 and ARM11.*

Table 2.1 Comparison Between Various ARM Families

Parameters	ARM7	ARM9	ARM10	ARM11
Pipeline length	Three –stage	Five-stage	Six-stage	Eight-stage
Architecture	Von-Neumann	Harvard	Harvard	Harvard
MIPS/MHz	*0.97*	*1.1*	*1.3*	*1.2*
Typical *MHz*	*80*	*150*	*260*	*335*
Multiplier supported	*8 × 32*	*8 × 32*	*16 × 32*	*16 × 32*

2.9 SUMMARY

In this chapter, evolution of *ARM* processors, *ARM* as a modified *RISC* architecture with its additional features and data flow model of typical *ARM* processor are discussed in detail. Load-Store architecture, *ARM's* programmer's model and *ARM* instruction set are discussed in brief. The chapter also discusses pipeline behavior of *RISC* processor by considering execution of data processing instruction. Finally, the chapter ends with comparison of various *ARM* processors families.

CHAPTER 3

ARM CORTEX– M3

3.1 THUMB -2 TECHNOLOGY

As discussed in the previous chapter, the *ARM* instruction set provides the definitive and complete *32-bit* instructions for *ARM* architecture while Thumb instruction set is an extension to the *32-bit ARM* architecture that enables high code density. The Thumb instruction set provides a subset of most commonly used 32-bit *ARM* instruction set, which have been encoded in to *16-bit* wide operation codes. During execution, these *16-bit* Thumb instructions are decoded to perform same function as *ARM 32-bit* instruction equivalents. An improvement in code density is achieved by Thumb instruction set but at the expense of performance. Single *16-bit* Thumb instruction is an exact equivalent to single *32-bit ARM* instruction and more *16-bit* Thumb instructions require to be executed to accomplish the same task. *ARM* instruction will show a performance advantage over Thumb instruction, when there is no difference in time taken to fetch an instruction from memory. Writing applications using Thumb instructions will enable most frequently used code to be stored in on-chip memory, which results in higher code density and application also compatible with the aim of achieving low power. When performance is the primary constraint, generally the usage of fewer instructions is a better approach. Hence, usage of *ARM* instructions alone will usually give the best results. In order to achieve the desired balance of *performance* and *code density* to produce an optimized design, designers tend to use a mixture of both *ARM* and Thumb instructions. However, there are some limitations in using this mixed *ARM*-Thumb approach. Since not all *ARM* instructions have Thumb equivalents, some *ARM* instructions must still be used even when the target is the highest code density. To access these, Thumb code must call an *ARM* code function, which performs the operation before returning to the Thumb code. These factors will decrease the performance of the core. In order to improve code density and performance of the core, Thumb-2 set was introduced. Thumb-2 core technology extends the *ARM* architecture to add enhancements to the Thumb *ISA* that benefit code density and performance [15].

Thumb-2 *ISA* consists of the existing 16-bit Thumb instructions augmented by:

- New *16-bit* Thumb instructions for improved program flow.
- New *32-bit* Thumb instructions derived from *ARM* instruction equivalents.
- At the same time the *ARM 32-bit ISA* has been improved with the addition of new *32-bit ARM* instructions for improved performance and data handling.

Fig. 3.1 shows the comparison between *ARM*, Thumb and Thumb-2 instructions sets with reference to code density.

Thumb can now access all of the instructions it needs to enable both high performance and code density. The new Thumb-2 core technology provides:

- Access to equivalents for virtually all *ARM* instructions.
- 12 completely new instructions, which change the performance-code size balance.
- As shown in Fig. 3.1, memory footprint is just *74%* of an *ARM*– equivalent implementation [15].
- Eliminating need of switching between *ARM* and Thumb stage enables faster speed of execution.

34

- Better than existing Thumb high density code[15] :
  - 5 percent smaller.
  - 2-3 percent faster.

Code Size

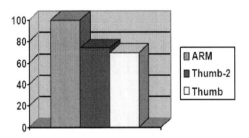

Fig. 3.1   Relative Code Density

With the development of Thumb-2 core technology, it has become the default instruction set for the majority of applications. Thumb-2 core technology is a significant enhancement to the *ARM* architecture, which provides performance at higher code densities than previously achievable with *ARM* architecture. Due to that all *ARM* Cortex profile micro controller i.e. *ARM Cortex- A series, ARM Cortex – R series, ARM Cortex -M* series has adopted *Thumb-2* core technology [14]. The Cortex processor families are the first products developed on architecture v7; the *ARM* extended core version. In this architecture, the design is divided into three different profiles as follows:

- *A Profile (ARMv7-A):*   Application processors required to run complex applications such as high-end embedded operating systems like Symbian, Linux, and Windows Embedded, requiring the highest processing power, virtual memory system support with Memory

Management Units (MMUs), and optionally, enhanced Java support as well as a secure program execution environment. Example products include high-end mobile phones and electronic wallets for financial transactions.

- *R Profile (ARMv7-R):* Real-time, high-performance processors targeted primarily at the higher end of the real-time market—those applications, such as high-end breaking systems and hard drive controllers, in which high processing power and high reliability are essential and low latency is also important.

- *M Profile (ARMv7-M):* Industrial control applications, including real-time control systems and processors targeting low-cost applications, where processing efficiency is important and cost, power consumption, low interrupt latency, and ease of use are critical.
.

In order to meet the requirements of the *32-bit* embedded processor market, followings are the different techniques used in the processor.

- Greater performance efficiency, allowing more work to be done without increasing the frequency or power requirements.

- Low power consumption, enabling longer battery life, especially critical in portable products including wireless networking applications.

- Enhanced determinism, guaranteeing that critical tasks and interrupts are serviced as quickly as possible but in a known number of cycles.

- Improved code density, ensuring that code fits in even the smallest memory footprints.

- Lower-cost solutions, reducing *32-bit*-based system costs close to those of *8-bit* legacy.

- Tools easily available for programming and debugging.

ARM has released first generation of Cortex processor termed as *ARM Cortex – M3*. It is an M- profile processor based on *ARM v7-M* architecture, primarily designed to target the *32-bit* microcontroller market. The *Cortex-M3* processor provides excellent performance at low gate count and comes with many new features previously available only in high-end processors. The *Cortex-M3* processor builds on the success of the *ARM7* processor to deliver devices that are significantly easier to program and debug and yet deliver a higher processing capability.

3.2 ARM CORTEX– M3 PROCESSOR FEATURES AND APPLICATIONS

3.2.1 Features

The *Cortex- M3* processor is a *32-bit RISC* processor featuring low gate count, low interrupt latency, low power and low-cost debug. It is primarily developed for deeply embedded applications that require fast interrupt response features.

The processor incorporates followings:

Processor core: The processor core implements the *ARM v7-M* architecture. It has the following main features [17]:

- Thumb-2 (ISA) subset consisting of all base Thumb-2 instructions, *16-bit and 32-bit*: For high code density.
- Harvard processor architecture: For enabling simultaneous instruction fetch with data load/store.
- Three-stage pipeline: Smaller chip size.
- Single cycle *32-bit* multiply: Improves execution speed.

- Handler and Thread modes: Differentiate normal and exception processing.
- Interruptible-continued *LDM/STM, PUSH/POP.*
- A separate Load Store Unit (*LSU*): decouples load and store operations from the Arithmetic and Logic Unit (*ALU*).
- A 3-word entry Pre-fetch Unit (*PFU*): Read the data in advance from memory.
- Nested Vectored Interrupt Controller (*NVIC*) [17]: It is closely integrated with the processor core to achieve low latency interrupt processing. It has the following main features:
  - External interrupts of *1 to 240* configurable sizes.
  - Bits of priority of *3 to 8* configurable size.
  - Processor state automatically saved on interrupt entry, and restored on interrupt exit, with no instruction overhead.
  - Priority grouping: This enables selection of pre-empting interrupt levels and non pre-empting interrupt levels.
  - Support for tail-chaining and late arrival of interrupts: This enables back-to-back interrupt processing without the overhead of state saving and restoration between interrupts.
  - Processor state automatically saved on interrupt entry, and restored on interrupt exit, with no instruction overhead.
  - Dynamic reprioritization of interrupts.

3.2.2 Processor Applications

With its low cost, high code density and small size, the *Cortex- M3* processor is widely used for following applications [14]:
- Automotive: Ideal application for the *Cortex- M3* processor is in the automotive industry because of very high-performance efficiency and

38

low interrupt latency, allowing it to be used in real-time systems. The *Cortex- M3* processor supports up to *240* external vectored interrupts, with a built-in interrupt controller to support nested interrupt and make it ideal for highly integrated and cost-sensitive automotive applications.

- Low-cost microcontrollers: The *Cortex- M3* processor is ideally suited for low-cost microcontrollers, which are commonly used in consumer products. Its lower power, high performance, and ease-of-use advantages enable embedded developers to migrate to 32-bit systems and develop products with the *ARM* architecture.

- Data communications: The processor's low power and high efficiency, coupled with Thumb-2 instructions for bit-field manipulation, make the *Cortex-M3* perfect for many communications applications, such as Bluetooth and Zig Bee.

- Industrial control: In industrial control applications, simplicity, fast response, and reliability are key factors. Again, the *Cortex-M3* processors interrupt feature, low interrupt latency and enhanced fault-handling features make it a strong processor in this area.

- Consumer products: In many consumer products, a high-performance microprocessor is used.

The *Cortex-M3* processor being a small processor, it is highly efficient and low in power consumption and supports an *MPU* enabling complex software to execute while providing robust memory protection.

3.3 PROCESSOR CORE ARCHITECTURE

To reduce the complexity and make the chip size smaller, the *Cortex- M3* core is based on 3-stage pipeline architecture [17]. Following stages make up the pipeline:

- Fetch Stage
- Decode Stage
- Execute Stage

Instruction Fetch/Prefetch (IF) Stage

In the first stage of pipeline architecture, requested instruction from memory is send to fetch stage via *32-bit I-Code* bus [17]. It is a *32-bit* bus based on the *AHB-Lite* bus protocol for instruction fetching in memory regions. Instruction fetches are performed in word size, even for Thumb instructions. Therefore, *CPU* core could fetch up to two Thumb instructions at a time during execution. Operation of the *IF* stage starts when the content of program counter (*PC*) is sent out to fetch instruction from memory and store in to instruction register (*IR*). The *IR* register is used to hold the instruction needed on subsequent clock cycles. As fetch stage consists of Pre-fetch Unit, which reads at one word /clock cycle from main memory through *I-CODE* bus. It consists of three words *FIFO* [17]. This can be two Thumb instructions, one word-aligned Thumb-2 instruction, or the upper/lower half word of a half word-aligned Thumb-2 instruction with one Thumb instruction, or the lower/upper half word of another half word-aligned Thumb-2 instruction [17].

All fetch addresses from the core are word aligned. If a Thumb-2 instruction is half word aligned, two fetches are necessary to fetch the Thumb-2 instruction. However, the 3-entry pre-fetch buffer ensures that a stall cycle is only necessary for the first half word of Thumb-2 instruction fetched. Pre-fetching is a common mechanism in pipeline architecture and use to increase strength of the core by reading data well in advance from memory.

Data forwarding is done from pre-fetcher to fetch stage in the same clock cycle. Fetch stage consists of address computation unit that sequentially increments address with +4 in the case of sequential execution. When a branch instruction is encountered, the decode stage also includes a speculative instruction fetch that could lead to faster execution. The processor fetches the branch destination instruction during the decode stage itself. Later during the execution stage, the branch is resolved and the instruction to be executed next is known. If the branch is not to be taken, the next sequential instruction is already available. If the branch is to be taken, the branch instruction is made available at the same time as the decision is made restricting idle time to just one cycle. In case of a conditional branch instruction decoded, required control signals are generated and address computation unit add the current value of PC with offset address specified as a part of instruction and pre-fetch the branch target address instruction called *branch forwarding*, which is used to reduce one of wait state.

When running programs with mostly *16-bit* instructions, processor may not fetch instructions in every cycle. This is because the processor fetches up to two instructions (*32-bit*) in one go. So, after one instruction is fetched, the next one is already inside the processor. In this case, the processor bus interface may try to fetch the instruction after the next or, if the buffer is full, the bus interface could be idle. Some of the instructions take multiple cycles to execute. The pipeline will be stalled in such conditions. Note that due to the pipeline nature of the processor and to ensure that the program's compatibility with Thumb codes, when the program counter is read during instruction execution the read value will be the address of the instruction plus 4. This offset is constant and independent of the combination of *16-bit* Thumb instructions and *32-bit* Thumb-2 instructions. This ensures consistency between Thumb and Thumb-2 [14].

Instruction Decode Stage

In *RISC* architecture as op-codes are fixed in size, it is designed as per number of bits occupied by operands and the remaining bits were used to determine various categories of instructions. In this stage, instruction decoder gets the instruction in form of word data from the fetch stage, which is decoded and necessary control signals are generated to perform the required operation in later part of the clock cycle. By decoding the content of *bits [15:11]*, either *16-bit or 32-bit* instructions are differentiated and subsequent operations are performed in the later part of the clock cycle [22]. After differentiating *16* and *32-bit* instructions, various instruction categories like data processing with shift register, data processing with immediate values, load with multiple registers, conditional branch, unconditional branch etc. are identified and function codes are generated with necessary control signals. Function codes are used to instruct subsequent units to carry out further operation as per pipeline behavior. In the same clock cycle *register file* is accessed to perform the required register read operation [15]. Note that, in case of decoding two *16-bit* instructions, one of the instructions is stored in internal buffer of the decoder and pre-fetching operation is suspended for one clock cycle.

However, a separate address computation unit is used for Load/Store related instruction in this stage to restrict the architecture up to three stage pipeline [15]. Function of address computation unit is to compute the memory address by adding the content of base register with offset as specified as a part of instruction. From this address, data are transferred to/from memory as per *load/store* operation.

Instruction execution stage

From the instruction decoder stage, the operands, function code and necessary control signals are sent to execution stage. The execution stage performs operations on operands as per function code received from instruction decoder stage. Followings are the various operations carried out in this stage:

- Shifting operation on *operand 2* as per shift count (optional).
- Data processing operation on *operand 1* and *operand 2* as per function code received from the decoder stage.
- Load-Store data access operation, if required.
- Register-write operation to write back result in to register file.

Components of this stage are barrel shifter, *Load/Store* data access unit, *ALU* and hardwired multiplier as explained below.

Barrel shifter

Processor combines a shift operation in every data processing instruction at no extra cost in terms of code and little additional cost in terms of time. Barrel shifter is a purely combinational device that performs multiple bit shifts in same clock cycle (execution stage). It is different from conventional shifter, which can perform one bit shift on one clock cycle, while barrel shifter is able to produce multiple bits shift in the same clock cycle. In addition, shifter can generate constants in the range of *0 to 31* and is also used to perform memory address offset calculation for load/store related instructions. Shift data is an immediate data or register value depending upon execution of instruction. Shifter performs various shift operations as follows:

- *LSL*: logical shift left by *0 to 31* places; fill the vacated bits at the least significant end of the word with zeros.
- *LSR*: logical shift right by *0 to 32* places; fill the vacated bits at the most significant end of the word with zeros.
- *ASL*: arithmetic shift left. This is a synonym for *LSL*.

- *ASR*: arithmetic shift right by *0 to 32* places; fill the vacated bits at the most significant end of the word with zeros if the source operand was positive, or with ones if the source operand was negative.
- *ROR*: rotate right by *0 to 32* places; the bits which fall off the least significant end of the word are used, in order to fill the vacated bits at the most significant end of the word.
- *RRX*: rotate right with one bit extended.

Processor preprocesses *operand 2* by carrying out shifting operation as per the specified shift count and sends the operand to *ALU* stage for further processing.

Load-Store data access Unit

Function of this unit is to transfer data to/from memory as per load or store operation. *Cortex- M3* processor incorporates Harvard architecture to speed up the execution [17]. Separate *D-Code* bus is used to transfer data to/from memory. Hence, *I-Code* bus continues to fetch the next instruction from memory simultaneously increasing the speed of execution greatly. *D-Code* bus is a *32-bit* bus based on the *AHB-Lite* bus protocol. It is used for data access in memory regions during load-store operations. Load operation with multiple registers can load the data into specific registers via *D-Code* bus. Store operation with multiple registers can store the data into specified memory location. Any of the *ARM's* 16 registers can act as an address (i.e. index) register. *ARM* architecture refers register indirect addressing mode as an *indexed addressing*. Load instructions require two cycles for first access and one cycle for each additional access [17]. For an example, Load instruction with two registers require three cycles; two cycles for first access and one cycle for loading the content of second register.

Arithmetic Logical Unit (ALU)

The unit is responsible for processing operands as per the received function code and computing the result. *ALU* does not only add its two inputs but also performs full set of data operations defined by the instruction set, including branch calculations, bit-wise logical functions, and so on. The unit receives operands from decoder unit and shifter unit, receives function code from decoder stage, process on them and computes the result/address. After that *register-write* operation is initiated to write the result in destination register. This unit accesses the register file for reading content of Program Status Register (*PSR*) to obtain the previous value of conditional flags. In the case of execution of unconditional branch instruction, the unit adds the current value of program counter with the offset specified as a part of operand and computes the branch instruction target address. The address is sent to fetch stage address computation unit for fetching the branch target instruction.

Booth's Multiplier

All *ARM* processors apart from the first prototype have included hardware support for integer multiplication. Two styles of multiplier have been used:

- Older *ARM* cores include low-cost multiplication hardware that supports only the *32-bit* result multiply and multiply-accumulate instructions.

- Recent *ARM* cores have high-performance multiplication hardware and support the *64-bit* result multiply and multiply-accumulate instructions.

The multiplier employs a modified Booth's algorithm to produce the *2-bit* product, exploiting the fact that x3 can be implemented as x (-1) + x4. This allows all four values of the *2-bit* multiplier to be implemented by a simple shift and add or subtract, possibly carrying the x 4 over to the next cycle.

3.4 ARM CORTEX- M3 PROGRAMMER'S MODEL

As discussed in the section 3.1, the *Cortex– M3* processor implements the *ARM v7-M* architecture. This includes the entire *16-bit* Thumb instruction set and the Thumb-2 *32-bit* instruction set architecture. The processor cannot execute *ARM* instructions. Thumb instruction set is a subset of the *ARM* instruction set, re-encoded to *16 bits*. It supports higher code density and systems with memory data buses that are *16 bits* wide or narrower. Thumb-2 is a major enhancement to the Thumb Instruction Set Architecture (*ISA*). Thumb-2 enables higher code density than Thumb and offers higher performance with *16/32-bit* instructions.

Operating modes

The processor supports two modes of operation, *Thread mode* and *Handler mode*:

- *Thread* mode is entered on Reset, and can be entered as a result of an exception return. *Privileged* and *User* (Unprivileged) code can run in Thread mode.
- *Handler mode* is entered as a result of an exception. A code is privileged in *Handler* mode.

Registers

The processor has the following *32-bit* registers:

- *13* general-purpose registers, *r0-r12*.
- Stack point alias of banked registers, *SP_process* and SP_main.
- *Link* register, *r14*.

- Program counter, *r15*.

- One program status register, *xPSR*.

General Purpose Registers (*GPRs*): The register *r0 to r12* are defined as General Purpose Registers, which have no special architecture-defined uses. The *GPRs* are divided into following categories:

Low registers: Registers *r0* to *r7* are termed as low level registers, which are accessible by any instructions.

High registers: Registers *r8* to *r12* are termed as high level registers, which are accessible by all *32-bit* instructions. These registers are not accessible by all *16-bit* instructions.

The registers *r13, r14, r15* have the following special function:

Stack Pointer (*SP*): The register *r13* is used as the stack pointer (*SP*). The lowest two bits of the stack pointers are always *0*, which means they are always word aligned.

The *Cortex-M3* contains two stack pointers, *R13*. They are banked and only one is visible at a time:

- Main Stack Pointer (*MSP*): The default stack pointer; used by the OS kernel and exception handlers.

- Process Stack Pointer (*PSP*): Used by user application code.

Link Register (*LR*): Register *r14* is the subroutine Link Register (*LR*). When a subroutine is called, the return address is stored in the link register. The *LR* receives the return address from *PC* when a Branch and Link (*BL*) or Branch and Link with Exchange (*BLX*) instruction are executed. The *LR* is also used for exception return.

Program Counter (*PC*): Register *r15* is the *Program Counter* (*PC*). The *Program Counter* contains the current program address. This register can be written to control the program flow. *Bit [0]* is always 0. Thus, instructions are always aligned to word or half word boundaries.

Special function Program Status Register (x PSR):
Special registers in the Cortex-M3 processor include:

- Program Status Registers (*PSRs*)
- Interrupt Mask Registers (*PRIMASK, FAULTMASK, and ASEPRI*)
- Control Register (*Control*)

The program status registers are subdivided into three status registers: (i) Application PSR (*APSR*) (ii) Interrupt PSR (*IPSR*) (iii) Execution PSR (*EPSR*). Interrupt Mask Registers (*PRIMASK, FAULTMASK, and BASEPRI*) are used to disable exception in timing – critical tasks. Control Register is used to define the privilege level and stack pointer selection. Combination of all three *PSRs* is termed as *xPSR*.

Table 3.1 shows the use of each bit field in Program Status Register (*PSR*).

Table 3.1 Use of Bit Field in Program Status Register

Bit	Description
N	Negative
Z	Zero
C	Carry/borrow
V	Overflow
Q	Sticky saturation flag
ICI/IT	Interrupt-Continuable Instruction (ICI) bits, IF-THEN instruction status bit
T	Thumb state always 1.
Exception Number	Indicates exception number.

48

Instruction Set

As discussed in section 3.1, the *Cortex –M3* processor supports the Thumb-2 Instructions set. This is one of the important features of *Cortex –M3* processor because it allows *32-bit* instruction and *16-bit* instructions to be used together for high code density and high efficiency. Since there is no need to switch between states, *Cortex- M3* processor has a number of advantages over conventional *ARM* processors. Some of them are listed below:

- No state switching overhead, saving both execution time and instructions space.
- No need for separate *ARM* and Thumb code source file, making software foundation easier.
- Easier to get the best efficiency and performance.

Instruction set consists of *16-bit* and *32-bit* Thumb instructions. Bits *15 to 10* are used to differentiate the various instruction categories and remaining bits are used to store the operands.

32-bit Thumb instruction encoding and important categories:

In the *32-bit* encoding, position of bits *15:13*, bits *12* and *11*, bits *10:4* are fixed in first half word. In second half word, position of *bit 15* is fixed. These bits are used to separate out various instruction categories.

Instruction categories

- Load/Store multiple.
- Data processing (Shifted register).
- Data processing (Shifted immediate).
- Data processing (register).
- Multiply, and multiply accumulate.

49

3.5 ARM CORTEX – M3 BUS INTERFACE:

Advanced Microcontroller Bus Architecture (*AMBA*) specification defines an on-chip communication standard for designing high performance embedded microcontrollers. Three distinct buses are defined with the *AMBA* specification [20]:

- The Advanced High- performance Bus (*AHB*).
- The Advanced System Bus (*ASB*).
- The Advanced Peripheral Bus (*APB*).

The Advanced High- performance Bus (*AHB*) is a new generation of *AMBA* bus. It is high performance system bus that supports multiple bus masters and provides high- bandwidth operation.

The Advanced System Bus (*ASB*) is a first generation of *AMBA* bus. A typical *AMBA ASB* system contains one or more bus masters i.e. Direct Memory Access (*DMA*) or Digital Signal Processor (*DSP*) are the common bus master in *ASB* bus.

The Advanced Peripheral Bus (*APB*) is used to interface to any peripherals, which are low bandwidth and do not require high performance of a pipelined bus interface.

Note that the bus interface on the *Cortex – M3* processor core is based on *AHB-Lite* bus protocols [17]. Features, control signals and operation of *AHB* bus are discussed in the following subsections.

3.5.1 Introduction to Advanced High Performance Bus (AHB)

It is a high-performance system bus that supports multiple bus masters and provides high-bandwidth operation. The *AMBA AHB* implements the features required for high-performance, high clock frequency system including [20]:

- Burst transfers
- Split transactions
- Single- cycle bus master handover
- Non_- tri state implementation
- Wider data bus configuration_(*64/128 bits*)

As shown in Table.3.2, typical *AHB* system design contains the following control signals of master/slave for write/read bus transfer cycle [20]:

AHB Master: A bus master is able to initiate read and write operations by providing an address and control information on address bus. Only one bus master is allowed to actively use the bus at one time. In this project, we have used Pre-fetcher as a master device.

AHB Slave: A bus slave responds to read or write operations within a given address-space range. The bus slave responds back to the master the success, failure or waiting of the data transfer. In this project, we have considered memory as a slave device.

AHB arbiter: The bus arbiter ensures that only one bus master is allowed to initiate data transfers at a time. An *AHB* would include only one arbiter, although this would be trivial in single bus master systems. In this project, we have used only one Master (Pre-fetcher). There is no need to use AHB arbiter.

Table.3.2 AHB System Design with Various Control Signals

Name	Source	Description
HCLK	Clock source	This clock times all bus transfers. All signal timings are related to the rising edge of *HCLK.*
HADDR[31:0]	Master	The *32-bit* system addresses bus.
HTRANS[1:0]	Master	Indicates the type of the current transfer, which can be *NONSEQUENTIAL, SEQUENTIAL, IDLE or BUSY.*
HWRITE	Master	When *HIGH,* this signal indicates a write transfer and when *LOW,* a read transfer.
HSIZE[2:0]	Master	Indicates the size of the transfer, which is typically byte (8-bit), half word (16-bit) or word (32-bit).
HBRUST[3:0]	Master	Indicates if the transfer forms part of a burst. Four, eight and sixteen beat bursts are supported and the burst may be either incrementing or wrapping.
HWDATA[31:0]	Master	The write data bus is used to transfer data from the master to the bus slaves during write operations. A minimum data bus width of *32 bits* is recommended.
HRDATA[31:0]	Slave	The read data bus is used to transfer data from bus slaves to the bus master during read operations. A minimum data bus width of *32 bits* is recommended.
HREADY	Slave	When *HIGH,* the *HREADY* signal indicates that a transfer has finished on the bus. This signal may be driven *LOW* to extend a transfer.
HRESP[1:0]	Slave	The transfer response provides additional information on the status of a transfer. Four different responses are provided, *OKAY, ERROR, RETRY and SPLIT.*

3.5.2 Overview of AHB Operation:

A bus master starts an *AHB* read/write transfer by sending necessary address and control signals. These signals provide information about address, direction and width of the transfer, as well as an indication if the transfer forms part of a burst. Two different forms of burst transfers are allowed [17]:

- incrementing bursts, which do not wrap at address boundaries.
- Wrapping bursts, which wrap at particular address boundaries.

An *AHB* bus incorporates write and read data bus. A write data bus is used to move data from the master to a slave, while a read data bus is used to move data from a slave to the master.

In *AHB*, every transfer cycle consists of followings:

- The address phase consists of an address and control signals, which lasts only a single cycle and provides address and necessary control signals information.
- The data phase, which may require one or more cycle. The data transfer may be extended by using *HREDAY* signal.

Following operations take place while data transfer,

- The master must send the address the control signals onto the bus after the positive edge of clock cycle.
- The slave then samples the address and control information on next rising edge of the clock. Once the slave has sampled the address and control information, it provides the appropriate response within the same cycle. After finishing the transfer, the Slave device generates a high *HREADY* signal.
- For read transfer, Master device continuously checks the status of *HREADY* signal. If status of *HREADY* signal is found high then Master samples the data from the read data bus *HRDATA*.

The data phase can be extended using the *HREADY* signal. When *HREADY* signal is low, it causes wait states to be inserted into the transfer and provides extra time for the slave to sample the data.

Note that *AHB-Lite* bus is a single layer bus protocol that can support only single master with multiple slave devices. *I-Code and D-Code* buses are based on the *AHB-Lite* bus protocol used by the *Cortex – M3* processor in order to access instructions/data.

I-Code Bus

I-Code interface is a *32-bit* bus based on the *AHB –Lite* bus protocol [17]. The bus is used to fetch the instructions and vector from code memory space. Instruction fetches are performed in word size, even for Thumb instructions. Therefore, during program execution, the core fetches up to two thumb instructions at a time.

D-Code Bus

The *D-Code* bus is a *32-bit* bus based on the *AHB- Lite* bus protocol [17]. It is used for data access from code memory/*SRAM* memory region.

3.6   ARM CORTEX – M3 MEMORY SYSTEM:

Cortex – M3 processor has different memory architecture than traditional *ARM* processors [14]. The *Cortex- M3* has a predefined memory map that specifies which bus interface is to be used when a memory location is accessed. This feature also allows the processor design to optimize the access behavior when different devices are used. This grants the built-in peripherals, such as interrupt controller and debug component to be accessed by simple memory access instructions. Predefined memory map facilitates the *Cortex - M3* processor to be highly optimized for speed and ease of integration in a System- On-Chip (*SOC*) designs. The *Cortex- M3* design has an internal bus infrastructure optimized for memory usage. The design permits these regions to be used differently, i.e. data memory can be put into the *CODE* region, and program code can be executed from an external *RAM* region.

Followings are different memory regions:

Code memory region (0x00000000–0x1FFFFFFF): This region is executable region. We can also put data memory in this region. Data operations take place via the data bus interface.

SRAM memory region (0x20000000–0x3FFFFFFF): This region is intended for on-chip *SRAM*. Write is buffered, and the cache is Write-back and Write allocated. This region is executable. We can copy the code over here and execute it.

External RAM region (0x60000000–0x7FFFFFFF): This region is used either for on-chip or off-chip memory. The accesses are cacheable (*WB-WA*), and we can execute the code in this region.

3.7   SUMMARY

The chapter starts with evolution of Thumb-2 technology used in *ARM Cortex- M3* instruction set architecture. *ARM Cortex– M3* processor features and its advantages are discussed in brief. Main focus of the chapter is on *ARM Cortex–M3* processor core architecture and each of the blocks is described in detail. The processor programmer's model is also discussed in brief. Additionally, the chapter describes *Cortex– M3* bus interface with standard *AHB* bus protocol. Finally, the chapter ends with *Cortex– M3* memory system.

CHAPTER 4

MODELING OF ARM CORTEX– M3 ARCHITECTURE

This chapter describes the modeling of different architecture blocks of processor with various design challenges faced during modeling. We explain the behavior of different stages with the help of flow charts and pseudo codes.

4.1 MODELING OF INSTRUCTION FETCH/PREFETCH STAGE

Fig. 4.1 and Fig. 4.2 show the architecture block of instruction fetch stage and address computation unit of fetch stage respectively.

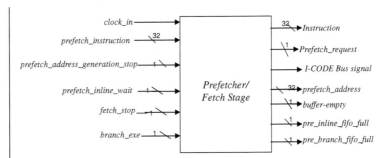

Fig. 4.1   Architecture of Fetch Stage

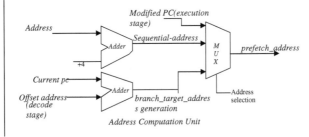

Fig. 4.2   Address Computation Unit of Fetch Stage

- As per the Fig.4.2 *prefetch_address* is generated from address computation unit and corresponding *prefetch_instruction* is received from memory in the next subsequent cycle through *I-CODE* Bus. Data forwarding is done in fetch stage in the same cycle.

- When conditional branch instruction is encountered at the decode stage, *prefetch_inline_wait* control signal is generated that stalls the sequential address generation and *branch_target_address* is computed in the decode stage. The computed address is sent to the fetch stage to fetch the branch target address instruction and stored into buffer.

- During execution stage, if branch condition is found true then branch target instruction is sent to the fetch stage else sequential execution continues. Note that, *3-word* buffers are required in Prefetcher Unit (*PFU*) as per the specifications [17].

4.2 MODELING OF INSTRUCTION DECODES STAGE

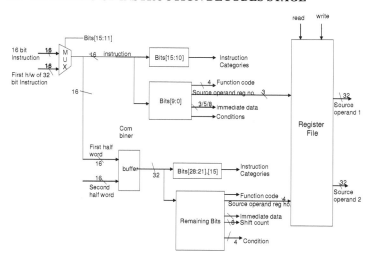

Fig. 4.3 Architecture of Instruction Decode Stage

57

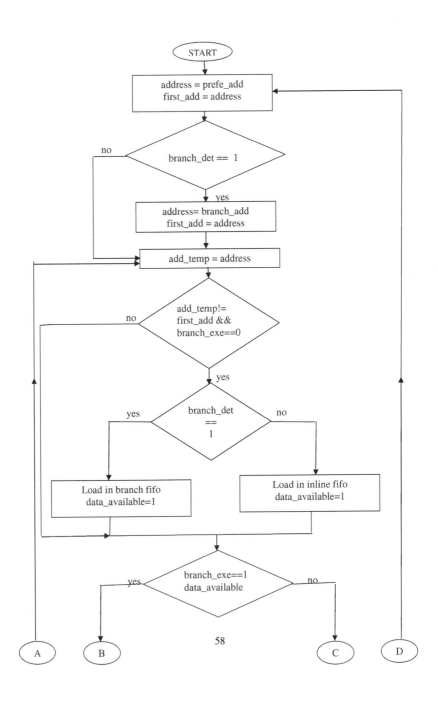

58

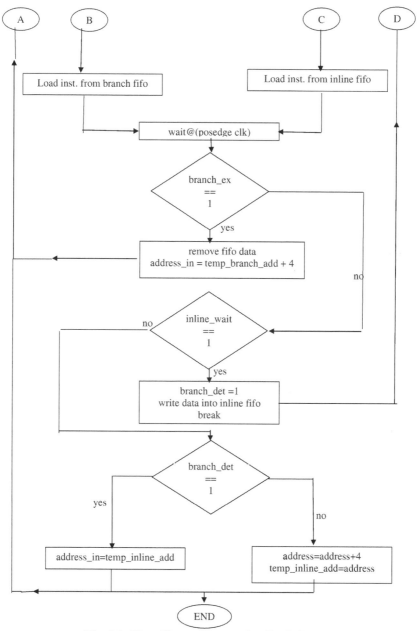

Fig. 4.4 Flow Chart of Instruction Fetch Stage

- As per Fig. 4.3, word data coming from fetch stage consists of one *16-bit* instruction and first half word of *32-bit* instruction. *Bits [15:11]* are used to differentiate 16-bit or 32-bit instruction types.

- If instruction is a *16-bit* instruction, *bits [15:10]* determine various categories of *16-bit* instructions like data processing (register), data processing (immediate), conditional branch etc and *bits [9:0]* are used to determine various function codes, source operand registers number, immediate data and their conditions related to branch instruction.

- If instruction is found to be *32-bit* instruction then first half word of current word data and second half word from subsequent cycle are combined in buffer to make the *32-bit* instruction. *Bits[28:21]* and *bit[15]* of *32-bit* instruction determine various instruction categories and remaining bits are used to determine various function code, source operand registers, immediate data, shift count and conditions related to branch instruction. Subsequently, read control signal is generated to read the content of source operand register(s),which store different operands, *source operand-1* and *source operand-2*.

- Note that the *Register file* requires three source operand read ports and two write ports for reading the value of source operands to initiate register write operation respectively. All three independent read operations and two independent write operations are performed in single clock cycle. So, all instructions can collect their source operands in single clock cycle except some of the instructions like multiple load/store instructions, multiply instruction etc.

- Among three independent read port, first read port is activated by decoder unit to access content of first source register *Rm* and second source register *Rn*. The second read port is activated by Load/Store unit to read the content of multiple registers during store operation and third read port is activated by *ALU* unit to read the content of

60

Program Status Register (*PSR*) register in order to read the previous value of conditional flags. Among two independent write ports, first write port is used by *ALU* to update value of destination register and second write port is used by Load/Store Unit during load operation to read the values from multiple registers and store into sequential memory stage. Flow chart of *Thumb* or *Thumb-2* instruction selection logic and functionality of Instruction decoder with control unit are as shown in Fig. 4.5 and Fig 4.6 respectively.

- *ARM v7- M* architecture decoding requirements vary according to instruction size because the architecture consists of 16-bit Thumb instructions and *32-bit* Thumb-2 instructions.

- As shown in Fig.4.5, word data (*32-bit*) comes from fetch stage to decode stage for decoding operation. The word data consists of two half words (*16-bits*) with the following different possibilities.

  (i)   Two consecutive *16-bit* instructions.

  (ii) One complete *32-bit* instruction.

  (iii) *16-bit* instruction followed by first *H/W* of next *32-bit* instruction.

  (iv) Second *H/W* of previous *32-bit* instruction followed by *16-bit* Thumb instruction.

  (v)   Second *H/W* of *32-bit* instruction followed by first *H/W* of next *32-bit* instruction.

- At the positive edge of every clock cycle, instruction selection logic first checks the status of *fetch_wait* signal, which is set during decoding of previous instruction. If the *fetch_wait* signal is low then the received word data is divided into two separate half-words (H/W), *Inst1* and *Inst2* respectively.

61

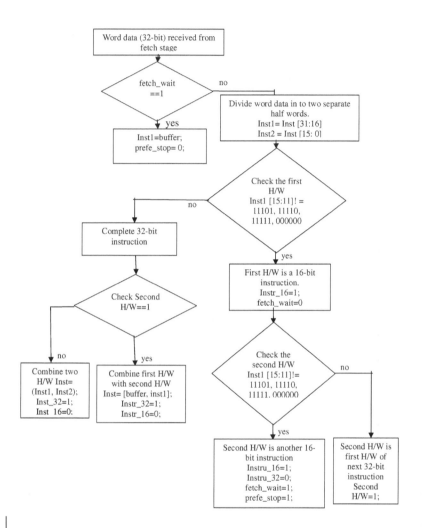

Fig. 4.5   Flow Chart of Thumb or Thumb-2 Instruction Selection Logic

- The algorithm then checks the first *H/W* for the value *11101*, *11110*, *11111* and *00000*. If the first *H/W* does not match with any of the above mentioned value then first *H/W* must be a *16-bit* Thumb instruction. Consequently, instruction selection logic generates

62

instru_16 signal for further decoding operation.

- The algorithm then checks the status of second *H/W* for the value *11101, 11110, 11111* and *00000*. If the second *H/W* does not match with any of the mentioned value then the second *H/W* must be the next consecutive *16-bit* Thumb instruction and required to be stored in the internal buffer of decoder. In response of this, *instru_16* signal is generated with *fetch_wait* signal for further decoding operation. *fetch_wait* signal instructs the decoder that next instruction is already available in the internal buffer and there is no need to receive next word data. In any case, if the second *H/W* is matches with any of the above mentioned values then the second *H/W* must be the first *H/W* of the next *32-bit* instruction. In response to this, instruction selection logic generates *second_halfword* signal and stores the second *H/W* in to its internal buffer.

- In case the first *H/W* matches with one of the above mentioned values, it indicates that the incoming word data is a complete or a part of a *32-bit* instruction.

- The algorithm then checks the status of *second_halfword* control signal, if it is found *HIGH* then combines first *H/W* (previously stored) with the second *H/W* (Inst1) and generate the signal *instru_32* for further decoding operation. If *second_halfword* is low then algorithm combines both the current H/W and makes one complete *32-bit* instruction and generates the *instu_32* signal for further decoding operation.

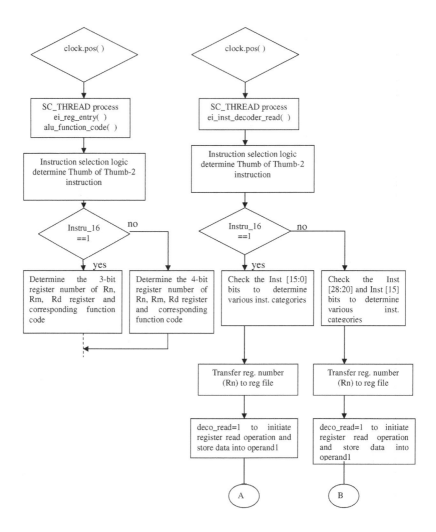

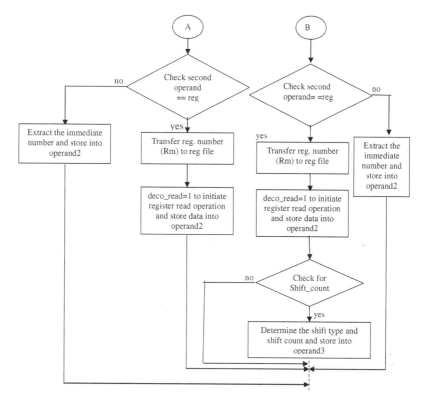

Fig. 4.6 Flow Chart of Instruction Decoder and Control Unit

The above Fig. 4.6 illustrates the flow chart of instruction decoder with control unit.

- As shown, the instruction decoder unit consists of *SC_THREAD* process *ei_reg_entry ( )* that is triggered at every positive edge of the clock cycle. As illustrated above in Fig. 4.5, Thumb or Thumb-2 instruction selection logic select either the control signal *Instru_16* or *Instru_32*.

- *ei_reg_entry( )* first checks the status of signal *Instru_16* and if it is found high then process determines the *3-bit* register number for register *Rn*, *Rm* and *Rd* from the received op-code. The process also determines the necessary *alu_function_code* for the relevant instruction.

- If signal *Instru_16* is low then the incoming instruction must be a *32-bit* instruction. The process determines the 4-bit register number for registers *Rn, Rm, Ra and Rd,* as well as the corresponding *function_code* for the same.

- Control unit consists of a *SC_THREAD* process *ei_inst_deco_read( )* that is updated at every positive edge of the clock cycle. Thumb or Thumb-2 instruction selection logic generates either *Instr_16* or *Instru_32* signal as discussed above.

- If signal *Instru_16* is found high then process checks the status of *Inst [15:0]* bits to determine various instruction categories. If signal *Instru_16* is low then process checks the status of *Inst [28:20]* and *Inst [15]* to determine the various instruction categories. In both the cases, whether *Instru_16 or Instru_32* is high, the process transfers register number *Rn* to register file and generate *deco_read* signal to initiate the register read operation and store the value into the *operand1*. Then process checks the status of second operand, whether it is register source operand or immediate source operand. If it is a register source operand then transfer the register number *Rm* to register file and generate the *deco_read* signal to initiate the register read operation and store the value into the *opernad2*. If the second operand is immediate operand then algorithm extracts the immediate data bytes from received *op-code* and stores into the *operand2*. In case of decoding of *32-bit* instruction, process checks for the *shift_count* value. If the received op-code consists of *shift_count* then the algorithm extracts the specified shift type and shift count from the op-code itself.

4.3 MODELING OF INSTRUCTION EXECUTION STAGE

Instruction execution stage consists of Shifter and ALU. The architecture and S/W implementation of both the units are discussed in this section.

Shifter Unit

As shown in Fig. 4.7, shifter processes second source operand as per shift count, which is a part of either register or immediate data. Shift type is selected according to *Sel_Type* control signal in order to select one of the possible shift types *LSL, LSR, ASR, ROR.*

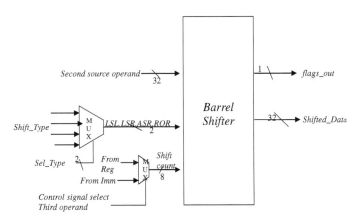

Fig. 4.7 Architecture of Shifter Unit

Shift count is also extracted by this unit, which is either a part of register or immediate. Shifter processes on second operand and produces *Shifted_Data* and *flags_out*. Shifter with *ASR* functionality is shown in Fig. 4.8.

Pseudo code: Showing *ROR* functionality.

If (shift type== ROR) then

If (Register Rm == shift count) then

shiftcount = bottom byte of Register Rm.

Dataout1 = copy the bits from Register Rn.

Dataout2 = >> on content of Rn as per shift count.

shifted_data = Dataout1 [shiftcount- 1, 0] & Dataout2.range [31-shiftcount, 0]

End If

End If

- As shown in Fig. 4.8, first the module determines necessary shift type i.e *LSR, ASR* etc. from the control signal *shift_type (2-bit)*. As the necessary shift type *ASR* is found, the module checks status of control signal *right_register_enable* or *right_imm_shift* in order to determine whether the shift count is a part of register or immediate data. If shift count is a part of register then only bottom byte of register is extracted.

- Subsequently in order to compute *ASR* shift type functionality, first LSR shift type is applied to *operand2* and result is copied in to one of temporary variables *temp_1*. Sign bit is copied as per number of shift count into variable *temp_2* afterwards. The shifted data is a combination of *temp_2* and *temp_1* in the range of bits *31 to 0*.

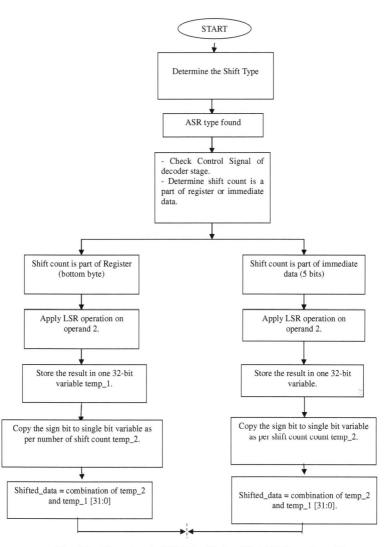

Fig. 4.8 Flowchart of Shifter Unit with ASR Functionality

Fig. 4.9 shows the architecture of Arithmetic Logic Unit (ALU). *ALU* is a purely combinational device that performs arithmetic and logical operations on *operand2* and *operand3,* which are received from register file and shifter unit respectively.

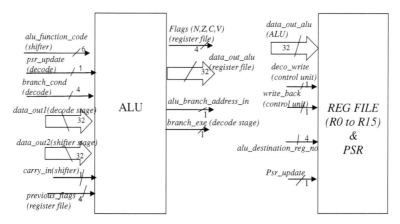

Fig. 4.9   Architecture of Arithemetic Logic Unit (ALU)

- *ALU* perfoms the desired operation according to   function code received from decoder/shifter stage. Signal *alu_function_code* (6-bit) provides internal mapping between decoder unit and *ALU* unit to instruct the *ALU* to carry out the desired operations.

- Note that the function code might be same for multiple instructions with different op-codes. In accordance with the received function code, *ALU* performs the desired operation on operands *data_out1* and *data_out2* and computes the result. For computation, the current value of flags must be referred from the decode stage.

- Further, ALU requires only single 32-bit adder with the look ahead

70

carry for execution of all arithmetic instructions. After performing computation, ALU checks the status of control signal *psr_update*. If status of the signal is high then it updates the desired flags (N, Z, C, V) as per requirements.

- Instruction execution timing is governed by the control unit (decode stage). Destination and PSR registers for corrosponding instructions are updated in register file by the control unit at the negative edge of same clock cycle. Control unit generates the *write_back* control signal as per cycle counter for the desired operation to update the destination and PSR registers.

- Note that, the destination register number is sent to register file during execution of current instruction. Register Address Register(RAR) holds the current destination register number untill current instruction reaches to execute stage.

- In case of execution of a conditional branch instruction, *ALU* first checks the status of *branch_code* (4-bit), which is received from decode stage. ALU checks the status of received flags as per the value of *branch_code*. If the status of the required flag matches with the value in *branch_code* then branch condition is found to be true and branching operation takes place. Subsequently, ALU generates *brach_exe* signal to instruct the decoder unit that brach execution is perfomed and normal prefetching operation required to be suspended. Execution starts from branch target address. Conditional and uncondtional branch instructions require common 32-bit adder in decode stage in order to compute the branch target address.

- In case of execution of conditional or unconditional branch instrucitons, pipeline is required to be reloaded. In pipeline reloading period, sequential instructions are executed but their destination registers are not updated. The destination register is only updated for branch target address instruction. If branch target address instruction is not executed then branch target address instruction required to be flushed from the prefetcher unit.

Fig 4.10 shows the flow chart of ALU by considering various control signals and processes.

- IE (Instruction Execution) register is a pipeline intermediate register used to store *operand1* and *operand2*, which are received from register file and shifter unit respectively. The ALU consists of *SC_THREAD* process *ei_alu_function ( )* that is triggered when module is getting *alu_flag* signal from the shifter unit.

- *alu_flag* signal indicates that operand2 is a valid shifted operand, which comes from the shifter unit. The algorithm then computes the desired operation on *operand1* and *operand2* as per the received function code from decoder and shifter unit. After computation, module checks the status of *psr_update*, which is received from the control unit. If status of this signal is found to be high then module will update the desired flags (*N, Z, C and V*) as per the operation.

- In the control unit, *SC_THREAD* process *ei_inst_time( )* is also running in parallel with *ei_alu_function( )* process. The process *ei_inst_time( )* is triggered at every positive edge of the clock and generates the *write_back* control signal as per the cycle counter for each of the operations. *alu_target_register* number is also sent to register file with the write back signal to enable the corresponding destination register. The process is used to generate the *PSR_reg* signal that enables the flag register of the register file.

72

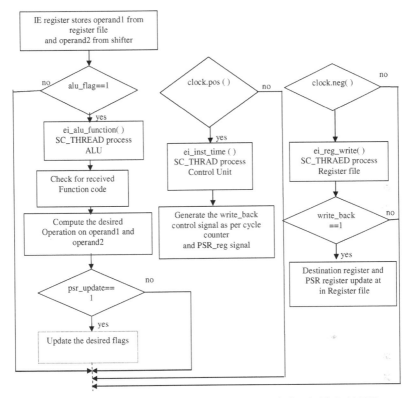

Fig. 4.10 Flow chart of Arithmetic Logic Unit (ALU)

- In Register file, *SC_THREAD* process *ei_reg_write( )* is simultaneously running in parallel with the above two processes and triggered at every negative edges of the clock. The process checks status of *write_back* control signal at every negative edge of the clock. If the status of the signal is high then corresponding destination and PSR registers are updated at the negative edge of the clock.

4.4 MODELING OF LOAD/STORE UNIT

RISC architecture consists of Load/Store unit in order to reduce the direct memory access using instructions. In load-store architecture, instructions that process data are operated only on registers and are separated from the instructions that access the memory. Operations either copy memory values into registers (load instruction) or copy register values into memory (store instruction). Note that, *Cortex – M3* processor incorporates the *Harvard* architecture to increase the speed of execution in which separate data bus is interfaced with the load-store unit to load or store the data from memory.

Fig. 4.11 shows the architecture of Load/Store unit. The architecture consists of Load/Store Address computation unit (decode stage) and Load/Store execution unit (execution stage).

- Load-store address computation unit consists of various control signals like $lsa\_mem\_acc\_in$, which is used to determine different addressing modes [8].

- The status of $lsa\_mem\_acc\_in$ with the value *00* indicates the pre-indexed addressing mode in which value of base register is added with the offset value in order to compute the transfer address. Transfer address is used to fetch the data from the desired memory location. The status of $lsa\_mem\_acc\_in$ with value *01* indicates the *post-indexed* addressing mode in which value of base register is used as the transfer address. This transfer address is used to fetch the data from the desired memory location. Control signal lsa_opcode_in determines function code for load-store operations i.e. load-store with single register, load/store with multiple register, load-store with base register update etc.

74

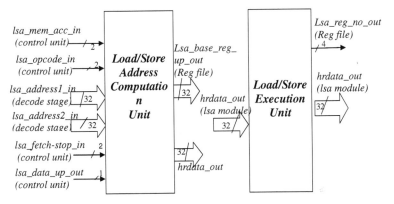

Fig. 4.11 Architecture of Load/Store Unit

- Note, the function code may be same for multiple instructions with different op-codes. Load/store address computation unit receives values *lsa_address1* and *lsa_address2* from decoder unit in order to determine the transfer address. *lsa_address1* is the value of the base register and *lsa_address2* indicates the offset value. Control signal *lsa_data_up_out* instructs the module, whether base register is required to update with the transfer address or it remains as it is. As per the status of *lsa_data_up_out* signal, *lsa_base_reg_up_out* (32-bit) is updated to further update the base register in the register file. *32-bit* bus *hrdata_out* sends the transfer address to the bus interface module.

- Load-store unit requires a separate bus interface module in order to interface with the external *AHB* bus. Load-store execution unit sends/receives data to/from memory using *32-bit* bidirectional data bus *hrdata_out*. In case of load operation, execution unit is also used to load the data into specified register R0 to R12 according to signal *lsa_reg_no_out* (4-bit).

75

Load-store architecture consists of load/store with multiple registers instructions [8]. These *LDM/STM* instructions can transfer subset of any *16* general purpose registers *R0* to *R15*. During execution of *LDM* instruction, control unit generates the number of *write_back* signals as per number of registers specified as a part of op-code. The module transfers data from lowest address into lowest register, which is specified as a part of op-code.

Fig. 4.12 illustrates the flow chart of Load/Store Addresses Computation unit.

- It consists of *SC_THRAED* process *lsa_add_mode_sel( )*. The process is triggered at every positive edge of the clock.

- The algorithm first checks status of *mem_acc* and if this signal is found *0* and the status of signal *fetch_stop_in* is low then process adds value of *base_address* with offset address and transfers it to *HADDR* bus in order to fetch the data from the memory. In this case, the base register is not updated and it preserves its old value. If *fetch_stop_in* signal is found high then process increments the address *HADDR= HADDR +4*.

- If the signal *mem_acc* is *1* and the status of signal *fetch_stop_in* is found to be low then process subtracts value of offset address from *base_address* and transfer it to *HADDR* bus to fetch the data from the memory. In this case also, the base register is not updated and preserves its old value. If *fetch_stop_in* signal is found to be high then process increments the address *HADDR= HADDR +4*.

- If the signal *mem_acc* is found with the value *2* and *fetch_stop_in* signal is low then process adds the value of *base_address* with the *offset address* (Pre_index addressing mode) and transfers it to the *HADDR* bus in order to fetch the data from the memory. In this case, base register is updated with the current value of *HADDR*. If *fetch_stop_in* signal is high then process increments the address *HADDR= HADDR +4*.

- If the signal *mem_acc* is found with the value *3* and *fetch_stop_in* signal is low then process transfers the *base_address* value to *HADDR bus* (Post_index addressing mode) to fetch the data from the memory location. In this case b*ase reg*ister is updated with the additio*n of base_add*ress value and the offset address value. If *fetch_stop_in signal* is *high* then process increments the address *HADDR = HADDR + 4*.

Fig 4.13 shows *the flo*w chart of load-store execution u*nit*.

- The algorithm consists of *ei_lsa_exe( ) SC_THREAD* pro*cess* that triggers at every positive edges of the clock. First, the algorithm determines the *s*tatus of controls signal *ld_st_single_cycle* in order to determine whether the received load-store instruction is a single register or multiple registers instruction. If instruction involves multiple register then algorithm determines the number of registers required to be load/store and accordingly the number of registers are updated with the desired value.

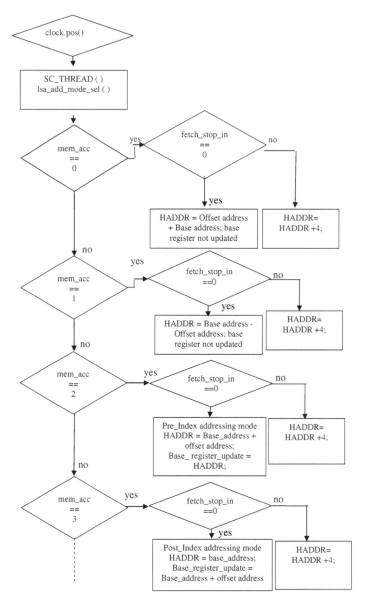

Fig. 4.12 Flow Chart of Load/Store Address Computation Unit (decode stage)

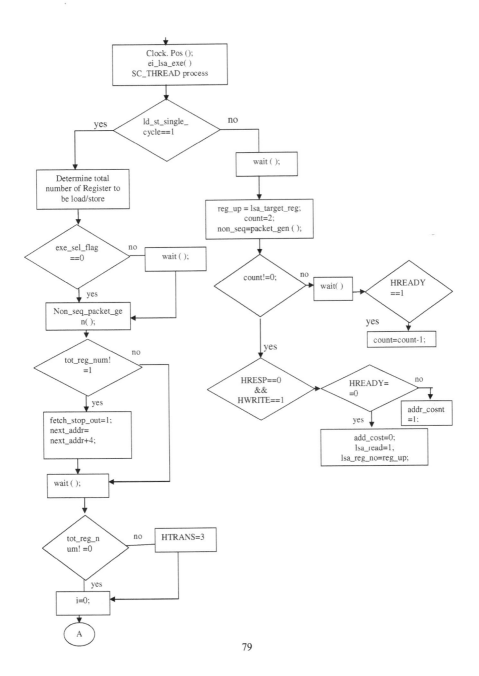

79

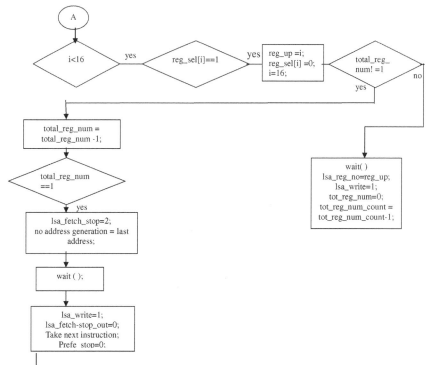

Fig 4.13 Flow Chart of Load/Store Execution Unit (execution stage)

4.5 MODELING OF ADVANCED HIGH PERFORMACE BUS

Advanced High Performance Bus (*AHB*) is a 32-bit external bus used to transfer op-code and data from the main memory within the given addressing space. *AHB* bus protocol contains one *AHB* Master and multiple *AHB* Slave devices. In this project, Pre-fetcher serves as a master device and memory serves as a slave device.

Architecture of *AHB* bus module is shown in Fig.4.14.

80

- The Master first checks the status of signal *address_gen_stop*. If this signal is found low then Master initiates read and write operations by transferring address and control information to the slave device. Various control signals are generated by master like *HWRITE* signal indicates write or read bus transfer. *HTRANS* control signal indicates whether the transfer consists of sequential addresses or non-sequential addresses.

- If status of signal *first_add_det* is high then master transfers *HTRANS* signal with value *10* to indicate the *NONSEQ* address else *HTRANS* signal with value *11* indicates *SEQ* address. Control signal *HSIZE* is provided to indicate whether transfer is a word, half-word or byte transfer. *HADDR and HWDATA* are 32-bit address and 32-bit write data bus used by the master to transfer address information and data to the slave device.

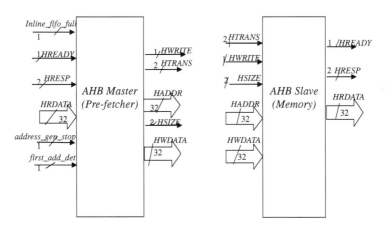

Fig 4.14   Architecture of Advance High Performance Bus (AHB)

- Slave device samples the address location by checking status of various control signals, which were generated by the master device

81

during previous clock cycle. As per status of received control signals, slave device performs the read/write operation. If status of *HWRITE* control signal is low then slave performs read transfer by transferring data on read data bus HRDATA else write transfer by writing data from HWDATA data bus to the selected memory location. Once the transfer has completed on data bus, slave device sets control signal *HREADY* high and also generates the control signal *HRESP* with the desired value that indicates additional information about status of a transfer.

- During read transfer Master device checks the status of *inline_fifo_full*. If this signal is found low then Master checks status of control signals *HREADY*, *HRESP* and *HWRITE*. If status of *HREADY* signal is found high with a low *HWRITE* then Master receives data from data bus *HRDATA* and increments the next sequential address else current address remains as it is.

Fig. 4.15 illustrates flow chart of AHB Bus transfer consisting of AHB Bus Master (Pre-fetcher) and AHB Bus slave (Memory).

- *AHB* Bus Master is a pre-fetcher unit that consists of *SC_THREAD* process *ei_fetch_memory( )*. This process is triggered at every positive edge of clock and first checks for the signal *address_generation_stop*. If the status of the signal is high then Master preserves the current PC value means next address is same as the current content of PC. If this signal is found low then Master checks the status of signal *first_add_det*. If this signal is high then Master initiates read bus transfer operation by sending control signals such as *HWRITE, HSIZE, HTRANS* and *HBRUST*. HTRANS control signal indicates that current transfer is NON_SEQUENTIAL transfer. *HBRUST* signal indicates incrementing burst of unspecified length.

- If the status of *first_add_det* signal is found to be low then Master initiates read transfer operation by sending control signal *HBRUST*ᴸ which indicates that current transfer is *SEQUENTIAL*.

- On the other side, AHB Bus consists of memory as a slave device consisting of SC_THREAD process *ei_ahb_slave_module( )*, which is triggered at every positive edge of clock. The process first checks the status of *HWRITE* signal. If this signal is low then process samples the address from memory array and transfers the data on read data bus *HRDATA*. The process generates the control signal *HREADY*ᵢ which indicates that current transfer has finished on the data bus.

- The process also generates *HRESP* signal that provides additional information about the status of a transfer. The Master on the other side continuously checks the status of *HREADY* signal and if this signal is high and *HRESP* signal contains the OKEAY response then Master extracts data from HRDATA data bus and stores into one of the buffers of Pre-fetcher unit. Consequently, Master increments the address value with +4 and sends to HADDR bus. If *HREADY* signal is found low then master provides the next address, which is same as current address value.

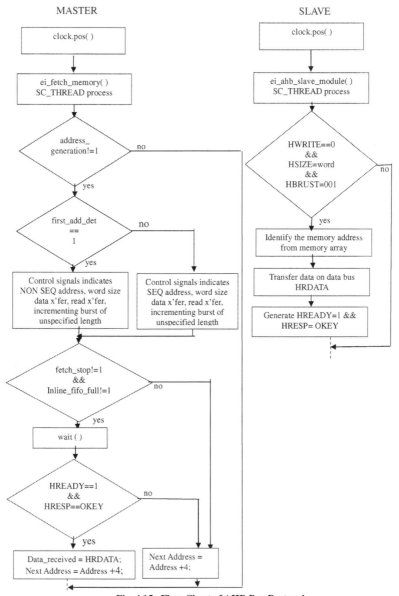

MASTER   SLAVE

clock.pos( )

ei_fetch_memory( )
SC_THREAD process

address_
generation!=1   no

first_add_det
==
1   no

Control signals indicates
NON SEQ address, word size
data x'fer, read x'fer,
incrementing burst of
unspecified length

Control signals indicates
SEQ address, word size
data x'fer, read x'fer,
incrementing burst of
unspecified length

fetch_stop!=1
&&
Inline_fifo_full!=1   no

wait ( )

HREADY==1
&&
HRESP==OKEY   no

Data_received = HRDATA;
Next Address = Address +4;

Next Address =
Address +4;

clock.pos( )

ei_ahb_slave_module( )
SC_THREAD process

HWRITE==0
&&
HSIZE=word
&&
HBRUST=001   no

Identify the memory address
from memory array

Transfer data on data bus
HRDATA

Generate HREADY=1 &&
HRESP= OKEY

Fig. 4.15   Flow Chart of AHB Bus Protocol

84

4.6 MODELING OF RESET EXCEPTION

Reset exception is a system exception. It has a fixed and highest priority in ARM processor architecture [26]. In case of assertion of Reset exception, vector address is required to be fetched from vector table in order to execute Reset Interrupt Service Routine (ISR). Vector table is an array of word data and used to determine the starting address of an exception handler. Vector table is initialized anywhere in memory but during Reset exception it is reset with the vector address *0*.

Fig. 4.16 shows modeling of Reset exception.

- Signal *Reset_exe* determines status of Reset exception. During Reset exception, vector table address is transferred to HADDR address bus with various control signals like *HTRANS*, *HSIZE* and *HWRITE* in order to initiate read bus transfer as discussed in the section 3.5.

- The module checks the status of control signals *HREADY* and *HRESP* in order to receive first, the initial value of SP_Main from HRDATA 32-bit read data bus. Once the initial value of SP_Main is received from data bus, the module sets control signal *write_reg* high. On receiving Reset vector address of Reset ISR from data bus, the module resets control signal *write_reg* low. Note that, the control signal *write_reg* is used by register file as shown in the following figure to update SP_Main or PC register with the desired value.

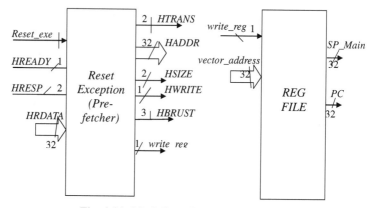

Fig. 4.16 Modeling of Reset Exception

Fig.4.17 shows the flow chart of Reset exception.

- As shown in flow chart Pre-fetcher module consists of *SC_THREAD* process *ei_fetch_memory( )* that is triggered at every positive edge of the clock cycle. The process first checks the status of Reset exception (*Reset_exe*) and if this signal is found high then process first computes the starting address of vector table by considering Vector Table Offset Register (VTOR) (29:7) bits. Content of Registers R0 to R12 is unknown during Reset exception. The process generates control signals like *HTRANS*, that indicates NONSEQUENTIAL *address* transfer *and* HWRITE signal that indicates read transfer operation.

86

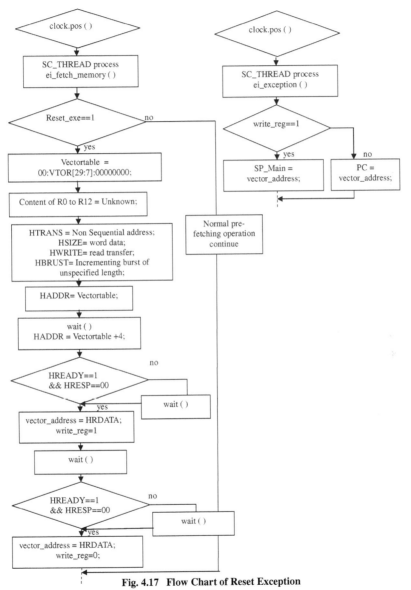

Fig. 4.17 Flow Chart of Reset Exception

87

- The algorithm then sends the address of vector table to HADDR bus and computes the *SP_Main* by logical *AND*ing the content of vector table address location with the *0xFFFFFFFC*. The process then increments vector table address location with +4 and determines value of Program Counter (*PC*). The process also sets the *Mode* signal, which indicates the current mode is Thread mode and *CONTROL* signal with value *00*, which indicates that current stack is Main and Thread is privileged. Reset exception process clears the current exception number and also the value of base priority register. If *Reset_exe* signal is low then normal pre-fetching operation is continued.

4.7 SUMMARY

This chapter discusses the major contribution of the dissertation. The chapter describes modeling of various architecture blocks of processor core like fetch stage, decode stage, execution stage and Load-Store unit. The chapter also describes modeling of AHB bus protocol and Reset exception. The chapter explains various design challenges faced during modeling. In this chapter, behavior of different stages of processor is explained with the help of flow charts and pseudo codes.

CHAPTER 5
SIMULATION & VERIFICATION

This chapter describes the environment used to simulate and debug the Design Under Test (*DUT*). The chapter also describes simulation results with various functionalities that describe behavior of important stages of the core.

5.1 SIMULATION ENVIRONMENT

To simulate the Design Under Test (*DUT*), *SystemC 2.2.0* version is used on a linux platform with kernel version of 2.6.18. *SystemC* set up includes library, which consists of source code for inbuilt available functions that can reduce the complexity of design. For an example, during design of pre-fetcher block of core, it requires 3- word *FIFO* [17], which is implemented by inbuilt function *sc_fifo* of SystemC. Another example is writing test cases for block *Register file*. It was required to pass random value through test case, which was implemented by *rand()* inbuilt function of *SystemC*. To compile and simulate the design following procedure is used in the project:

- By running the script file, Make file is executed with compilation commands and debugging option. An executable file is generated to simulate results.
- *sc_main( )* function consists of instances of test bench through which all design modules are instantiated. Design consists of multiple concurrent processes, which are invoked by clock signal or occurrence of events.
- Test cases are passed by writing test case name with script file.

89

- SystemC kernel dumps desired signals, which are traced in design and generate the VCD file to be viewed by GTKWave analyzer v3.2.2 for debugging purpose.
- Debugging is done using inbuilt function *sc_time_stamp( )* during execution.

5.2 SIMULATION RESULTS

This section describes the simulation results of different blocks of Cortex-M3 processor core such as instruction fetch stage, instruction decode stage, instruction execute stage and load-store architecture. This section also describes simulation results of *AHB* bus protocol and processor behavior during Reset exception. Simulations are carried out to validate the architectural model of *Cortex –M3* processor explained in Chapter 4. Stages of *ARM Cortex- M3* processor are simulated by passing various test conditions through test cases. The test cases are passed to *DUT* through test bench. Simulation results describe the behavior of individual blocks with reference to various design requirements. They also describe the behavior of individual blocks with reference to various timing relationships. The first stage of pipeline architecture, instruction fetch stage is discussed below with the simulation results.

Instruction Fetch Stage

Instruction fetch stage includes the functionalities like sequential address generation, prefetch/fetch instruction from memory and conditional branch detection as seen in Fig. 5.1. Prefetch_address *AAAAFFF0*H is sent to the pre-fetcher using a test vector *tb_pre_prefetch_address*. Data required to be stored into pre-fetcher are sent using the test vector *tb_pre_prefetch_instruction*. As seen in the figure, instruction fetch stage consists of two cycles for fetching first instruction from memory. In first cycle, 32-bit memory address is sent to the memory through I-Code Bus. This address is sequentially incremented in a word form (+4) by address incrementer i.e *AAAAFFF0H, AAAAFFF4H, AAAAFFFF8H*.

It can be seen that the address changes to *44444444H* from AAAAFFF8H, in step of four increments. This indicates detection of a conditional branch instruction at decode stage. However, the new address (*44444444H*) should be stored in fetch stage for facilitating the subsequent fetch operation from the new address. In fourth clock cycle, conditional branch instruction is encountered in the decoder stage and subsequently control signal *pre_dut_prefetch_inline_wait* is set. This signal stops sequential address generation and the branch target address instruction is stored in the buffer of Prefetcher instead of sequential instruction. Data is forwarded from Prefetcher to fetch stage in the same clock cycle. As seen in the figure, *pre_branch_exe* signal is low to indicate that condition of conditional branch instruction is false in execution stage and sequential address generation is initiated again.

Simulation results shown in Fig. 5.2 illustrate the situation when a conditional branch instruction is encountered in decoder stage and specified condition of conditional branch instruction is found to be true in execution stage.

As seen from Fig. 5.1, the condition of branch instruction is found true in execution stage to set the control signal *pre_branch_exe.*

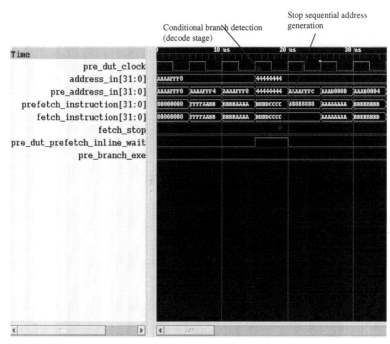

Fig. 5.1 Behavior of Fetch Stage for Detection of Conditional Branch Instruction

The signal *pre_branch_exe* causes Pre-fetcher to fetch the branch target address instruction i.e. 88888888H from memory and store into buffer of Pre-fetcher. It can be seen that data is forwarded from Pre-fetcher to fetch stage in the same clock cycle. In the next subsequent cycle, *pre_branch_exe* is reset that causes address incrementer to increment the address from branch target address i.e. *44444448H*, *4444444CH*, *44444450H*. Subsequently, data from these addresses are stored in prefetch stage i.e. *88888888H*, *99999999H*, *11111111H*. In the same clock cycle, they are forwarded to fetch stage. Note that, low *fetch_stop* signal indicates that all instructions require single clock cycle in execution stage.

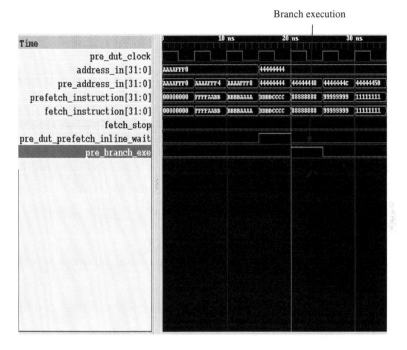

Fig. 5.2   Conditional Branch Detection with True Condition

Instruction decode stage

Function of instruction decode stage is to decode the op-codes, separate the operands and generate necessary control signals. Simulation result shown in below Fig. 5.3 illustrates decoding of *16-bit* data processing instruction category with immediate data as a second operand i.e. decoding of *ADD* instruction with *8-bit* immediate data. It can be seen that immediate data is taken with value *FFH*. Source and destination registers are same and use register *R1*. The collectively formed opcode of the *ADD* instruction *31FFH* is sent to decoder unit using test vector *tb_pre_prefetch_instruction*. Value

93

FFFF1111H is stored into register R1 using test vectors *reg_no, tb_data_in* and *tb_write*. In this stage, first the op-code is being decoded and register file is accessed to read the value of first source registers Rn i.e. *FFFF1111H*. After reading values from Register file, instruction decoder is used to separate out the operands data_out1 and data_out2 that store the values of first source register and immediate data i.e. *FFFF1111H* and 000000FFH respectively.

Necessary control signals generated such as *psr_update* indicate that Program Status Register (flags register) requires to update, *ALU_right_imm* and second operand is an immediate data. The stage also generates the function code *ALU_Type* i.e. *9* to instruct the execution unit to perform the subsequent ADD operation.

Simulation result shown below in Fig. 5.4 illustrates decoding of 16-bit *PC* relative address instruction *(ADR)*. As shown in the figure, the op-code formed is *4FF0H* by considering 8-bit immediate data with value *F0H*. The op-code of 16-Bit *ADR* instruction is transferred to decoder using test vector *tb_pre_prefetch_instruction*. Value *FFFF1111H* is stored into register *R15* using test vectors *reg_no, tb_data_in*.

As shown in below figure 5.3, in this case of instruction decoding, Program counter *(PC)* is read from register file which is a current value of *PC* i.e. *FFFF1111H*.

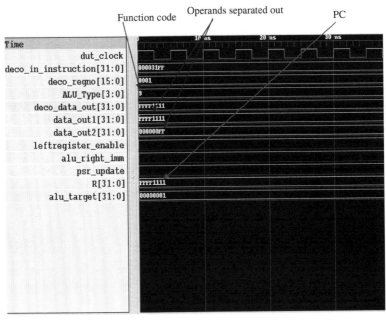

Fig. 5.3 Decoding of 16-bit Data Processing Instruction With Immediate Data

It is added with +4 and stored in to operand data_out1 i.e. *F0001115H* and 8-bit immediate data is stored in to operand data_out2 i.e. *000000F0H*. Necessary control signals such as *psr_update* generated with a low value indicates that flag register need not require to be updated. *ALU_right_imm* indicates immediate data in second operand. This stage also generates the function code *ALU_Type* i.e. *0* that instructs the execution unit to carry out subsequent operation.

95

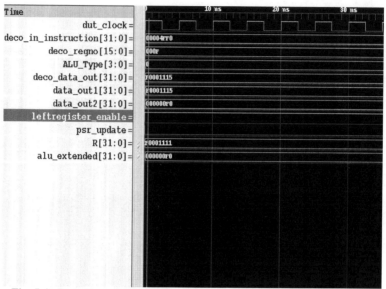

Time				
		10 ns	20 ns	30 ns
dut_clock =				
deco_in_instruction[31:0]=	0000EF0			
deco_regno[15:0]=	000F			
ALU_Type[3:0]=	0			
deco_data_out[31:0]=	F0001115			
data_out1[31:0]=	F0001115			
data_out2[31:0]=	000000F0			
leftregister_enable=				
psr_update=				
R[31:0]=	F0001111			
alu_extended[31:0]=	000000F0			

Fig. 5.4   Decoding of 16-bit PC Relative Address Instruction (ADR)

Simulation result shown in Fig. 5.5 illustrates decoding of 16-bit unconditional branch instruction. Opcode of the branch instruction *E7FEH* is sent to decoder unit using test vector *tb_pre_prefetch_instruction*. Value *F0001110H* is stored in to *R15* using test vectors *reg_no, tb_write, tb_data_in.*

During decoding of an instruction, first Program Counter is read from register file, which is current value of *PC* i.e. *F0001110H*. It is added with +4 and store in to operand *data_out1* i.e. *F0001114H*. Immediate data is separated and stored in to *data_out2* from *ALU*_extended i.e. *000007FEH*. Necessary control signals are generated to carry out further operations such as *psr_update* with low value indicating that flag register not required to update, *leftregister_enable* indicating that first operand is a part of register. It can be seen that *ALU_Type* signal indicates the function code, which

96

instructs the execution stage to carry out subsequent operations.

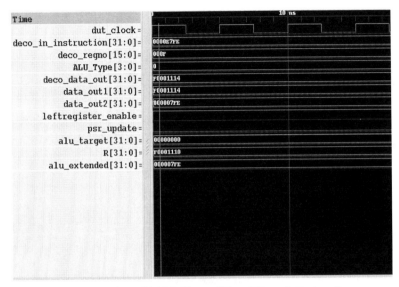

Fig. 5.5   Decoding of 16-bit Unconditional Branch Instruction

Shifter stage

Shifter is a combinational device that shifts multiple bits in same clock cycle. Processor uses shifter with no additional cost in terms of code but little additional cost in terms of time. The shifter is a part of execution stage.

Simulation results seen in Fig. 5.6 illustrate Logical Shift Right (*LSR*) operation of shifter. Shift type *01*, content of *operand2 F0001111H* and content of *operand3 F0000002H* are sent to shifter using test vectors *tb_shift_type* and *tb_data_in*. Shifter processes data of *operand2* as per specified shift count. S*hift_type 01* is coming from decoder stage and indicates Logical Shift right (*LSR*) operation. Control signal *in_rightregister_enable* indicates that shift count is a part of register i.e.

F0000002H. When shift count is a part of register it is always a bottom byte of register i.e. *02*. Shifting is perform on operand2 (in_data2) as per the specified *shift_count* that produced the shifted data *dataout_shift* i.e. 3C000444H. In *LSR* operation, all bits are shifted towards the most significant position according to shift count and all vacant bits are filled with zeros.

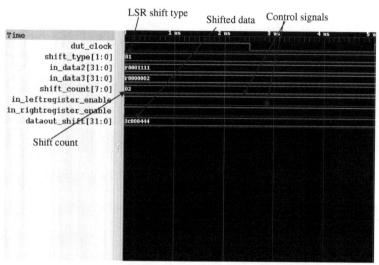

Fig. 5.6 Logical Shift Right (LSR) Operation of Shifter

Fig. 5.7 illustrates Rotate Right (*ROR*) operation of shifter stage. As shown in figure, shift-type *11* indicates the rotate right operation. As described for the previous result, here also control signal *in_rightregister_enable* indicates that shift count is bottom byte of register i.e. *09*. Rotate right operation takes place on *operand2* according to shift count and produces shifted data *dataout_shift* i.e. 88FFFF88H. In rotate right operation, shifted bits are wrapped around and combined with remaining bits.

98

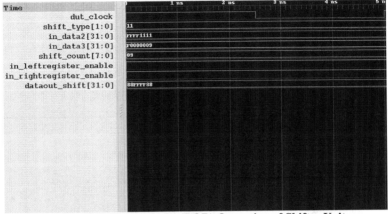

Fig. 5.7   Rotate Right (ROR) Operation of Shifter Unit

Integrated modules

Simulation results shown in Fig. 5.8 illustrate integrated result of modules such as prefetcher/fetch stage, instruction decoder stage and shifter unit. Pre-fetch address *AAAAFFF0H* with opcode of *32-bit AND* instruction EA110242H are transferred to pre-fetcher using test vectors *tb_pre_prefetch_address* and *tb_pre_prefetch_instruction*. Desired value is stored into registers R1 and R2 using test vectors *reg_no, tb_data_in*. In the first cycle, pre-fetch address i.e. *AAAAFFF0H* is sent to memory and corresponding instruction is received by pre-fetcher in the next clock cycle. In the same cycle, data is forwarded to fetch stage i.e. *EA110242H*. In the subsequent cycle, instruction is decoded in the decoder stage that will separate out operands *deco_data_out1, deco_data_out2* and *deco_data_out3* with the values *00000001H, 00000004H* and *00000001H* respectively. In this stage, various control signals are generated to carry out further operations in the next stage. In this stage also, necessary shift type is determined to instruct the shifter to perform necessary shift operation. It can be seen in figure that *00* shift type indicates Logical Shift Left (*LSL*)

99

operation required to be performed in execution stage. In the subsequent clock cycle, shift type is sent to shifter unit to shift *operand2* as per specified shift count i.e. *01* and produces shifted data *shift_dataout* i.e. *00000008H*. In the LSL operation all bits of operand 2 are shifted towards left significant position as per the shift count and all vacant bits are filled with zeros.

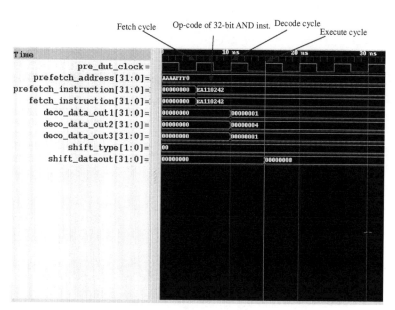

Fig. 5.8   Integrated Result of Modules Prefetch/Fetch, Instruction Decode and Shifter Stage with 32-bit AND Instruction

Fig. 5.9 shows the integrated results of designed module such as instruction fetch/pre-fetch stage, instruction decode stage and shifter unit. 32-bit TST instruction with Arithmetic Shift Right (ASR) forms the op-code *EA140FA5H*. In the first cycle, pre-fetch address i.e. *AAAAFFF0H* is sent to memory and corresponding instruction i.e. *EA140FA5H* is received by pre-fetcher in the next clock cycle. In the same cycle, data is forwarded into

100

fetch stage i.e. *EA140FA5H*. In the subsequent cycle, instruction is decoded in decoder stage that will separate out various operands i.e. d*eco_data_out1, deco_data_out2* and *deco_data_out3*. Different control signals are also generated in this stage. In this stage, shift type is determined to instruct the shifter to perform necessary shift operation. As shown in figure, shift type *10* indicates Arithmetic Shift Right (*ASR*) operation. In the subsequent cycle, shift type is sent to shifter unit, which shifts operand 2 as per the shift count i.e. *02* and produces the shifted data *shift_dataout* i.e. *E000000H*. In ASR operation, all bits of operand 2 are shifted towards most significant position specified as per shift count and vacant bits are shifted with the sign bit of *operand2*.

Thumb or Thumb-2 instruction selection logic:
Function of Thumb or Thumb-2 instruction selection logic is a part of decode logic used to select appropriate Thumb or Thumb-2 instruction (s) from incoming word (32-bit). ARM v7 – M architecture decoding requirements changes according to the size of instruction.

Simulation results shown in Fig 5.10 illustrate the Thumb or Thumb-2 instruction selection logic functionality. Pre-fetch address *AAAAFFFFH* and op-code of 16-bit *LSL* instruction followed by 16-bit LSR instructions are transferred to pre-fetcher using test vectors *tb_prefetch_address* and *tb_pre_prefetch_instrucion*. Then, 32-bit opcode consisting of first 16-bit LSL instruction followed by first half word of next 32-bit instruction (collectively formed as *4091EA13H*) is transferred to pre-fetcher.

The next 32-bit op-code consisting of second half word of previous 32-bit instruction followed by next 16-bit AND instruction (collectively formed as *00E433FFH*) is also transferred to pre-fetcher.

101

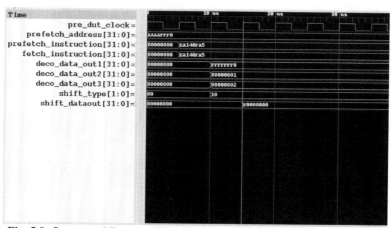

Fig. 5.9   Integrated Results of Prefetch/Fetch, Instruction Decode and Shifter Stage With 32-bit TST Instruction.

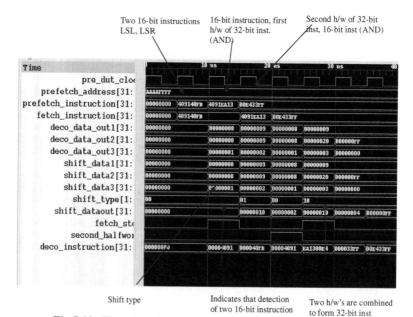

Fig 5.10   Thumb or Thumb-2 Instruction Selection Logic

As seen above, instruction selection logic decodes the first *16-bit LSL* instruction (*4091H*) and the next 16-bit LSR instruction (*40FBH*) is stored in the internal buffer of decoder unit. Operands are separated out for *LSL* instruction and necessary shift type is also determined during decode cycle. Note that, in the same cycle status of *fetch_stop* control signal is high, which indicates the module must not receive the next 32-bit data until the status of this signal is low. In the next clock cycle, instruction selection logic decodes 16-bit LSR instruction, which is already stored in internal buffer of decoder unit. In the same clock cycle, execution takes place for previous 16-bit LSL instruction and shifter produces the shifted data (*00000010H*). In the subsequent cycle, decoding is done for next 16-bit (*LSL*) instruction and signal *second_halfword* is high, which indicates that first half word of next 32-bit instruction is stored in the buffer of decoder. In the next clock cycle, instruction selection logic combines both half words in order to complete 32-bit AND instruction (*EA1300E4H*). Here, decoder unit receives both half words separately in two different clock cycles. In the same cycle, complete opcode is decoded for the 32-bit *AND* instruction and necessary shift type is also determined. The next cycle is the decoding cycle for the next 16-bit AND instruction.

Arithmetic Logic Unit (A LU) Unit

ALU is a purely combination device that performs arithmetic or logical operations on received operands. *ALU* updates desired flag as per the received function code from the decoder unit. Function code provides internal mapping between the decoder unit and the *ALU*; and instructs the *ALU* to carry out further operations. Function code may be same for multiple instructions with different opcodes.

Simulation results shown in Fig 5.11 illustrate execution of 16-bit add with carry (ADC) instruction functionality. Function code *05*, destination register

R7 and operands *7FFF0010H* and *6FFF0010H* are transferred to *ALU* using test vectors *tb_function_code*, *destination_register* and *tb_data_in*. As shown in simulation results, function code 05 instruct the *ALU* to perform arithmetic add with carry functionality. As per received function code, *ALU* performs addition of operand1 and operand2 with the previous stage carry. *ALU* receives *operand1* (*7FFF0010H*) and *operand2* (*6FFF0010H*) from register file and shifter unit respectively; and performs computation. Note that, during computation value of the previous stage carry is referred from the decode stage. After computation, *ALU* updates desired flags as per the received function code and subsequently PSR register is updated (*90000000H*) in register file at the negative edge of the same clock cycle. PSR register is updated by *PSR_update* control signal, which is generated from control unit. The control unit also generates register *write_back* signal as per the cycle counter for the corresponding operation and subsequently destination register R7 is updated (*EFFE0021H*) in register file at the negative edge of clock cycle.

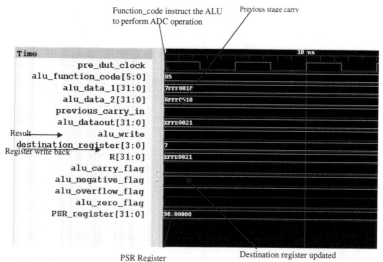

Fig 5.11 Execution of 16-bit Arithmetic (ADC instruction) Operation

104

Fig 5.12 shows simulation results of execution of 16-bit Logical (TST instruction) operation. Function code of TST instruction 00 is sent to *ALU* using test vector *tb_function_code*. Destination register R7 with operands FFFF0000H and 0000FFFFH are sent to *ALU* using test vector *tb_ALU_target_register1* and *tb_data_1* and *tb_data_2*. As per the received function code *00*, *ALU* performs logical bit-wise ANDing between two received operands (*FFFF0000H* and *0000FFFFH*) and computes the result. As the result after computation is zero, the corresponding zero flag is set by *ALU* and subsequently PSR register is updated in register file with the *value* (*40000000H*). As per the functionality of TST instruction, PSR register is required to be updated and result after computation must be discarded. As seen in figure, the destination register is not updated as *write_back* signal is set to low by control unit. So, destination register preserves its previous content but the PSR register is updated with the new value.

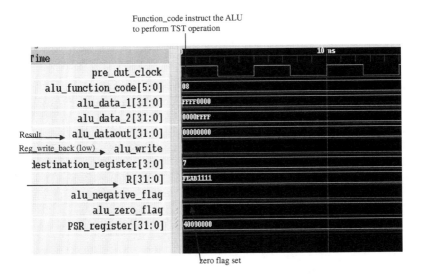

Fig 5.12 Execution of 16-bit Logical (TST instruction) Operation

Integrated modules

Fig. 5.13 illustrates results of the integrated modules such as prefetcher/fetch stage, instruction decoder stage, shifter unit and *ALU*. Pre_fetch address *AAAAFFFFH,* op-code of instruction 32-bit *BIC EA320343H* and op-code of instruction *32-bit MVN EA7F0445H* are transferred to pre-fetcher using test vector *tb_pre_prefetch_address* and *tb_pre_prefetch_instruction*. In the first cycle, pre-fetch address i.e. *AAAAFFFFH* and the subsequent address are sent to memory in order to pre-fetch the op-codes for the above mentioned instructions. In the second cycle, the module receives op-code for *BIC* instruction in the prefetcher unit and data is forwarded to the fetch stage at the negative edge of the same cycle. In the next cycle, decoding operation takes place for *BIC* instruction and also op-code for *MVN* instruction is received by the prefetcher and fetch stage as per the pipeline operation. In the decoding cycle, function code *0EH* is determined for *BIC* instruction and necessary control signals are generated to carry out further operations. The following cycle is execution cycle for BIC instruction in which first shifting operation is performed by shifter unit on *operand2* (*FA000000H*) as per the shift count *00000001H* and the shifted data *F4000000H* is produced. As per received function code *0E*, *ALU* performs *BIC* operation by bitwise *ANDing* of *operand1* with complement of shifted value and computes the result *0000A0FFH* in the same cycle. Destination register *R3* is updated by *write_back* control signal in execution cycle at the negative edge of clock cycle. *write_back* control signal is generated from control unit as per cycle counter for *BIC* operation. Because of pipeline operation, the same cycle is decoding cycle for the *32-bit MVN* instruction.

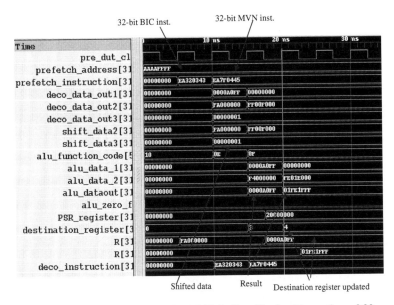

32-bit BIC inst. 32-bit MVN inst.

Shifted data   Result   Destination register updated

Fig 5.13   Integrated Result of ALU Unit Considering Execution of 32-bit BIC and 32-bit MVN Instruction

Next cycle is execution cycle of MVN instruction in which first shifting operation is performed on received operand2 as per the shift count *00000001H* and the shifted data *FE01E000H* is produced with the carry flag.

In the same cycle, *ALU* performs the *MVN* operations by bitwise inverse of shifted operand2 and computes the result *01FE1FFFH* as per received function code *0F*. At the negative edge of the same clock cycle, PSR register is updated with the value *20000000H*, which indicates that carry flag is set. Though MVN instruction is a logical instruction, it affects the carry flag because carry flag is set by the shifter unit during shifting operation. At the negative edge of the same clock cycle, destination register *R4* is updated

107

with the value *01FE1FFFH*. The destination register is updated by control signal *write_back*, which is generated from the control unit as per the cycle counter for the MVN operation.

Fig.5.14 shows simulation results of execution of 16-bit multiplication (*MUL*) instruction followed by 16-bit *AND* instruction. Pre-fetch address *AAAAFFFFH* and op-code for 16-bit *MUL* followed by opcode of 16-bit AND instruction *435D400AH* are transferred to the pre-fetcher/fetch stage using test vector *tb_pre_preftch_address* and *tb_pre_prefetch_instruction*. Consequently, op-code is received by pre-fetch/fetch stage. In the subsequent cycle, op-code is decoded for 16-bit *MUL* instruction. As discussed above, op-code for the next 16-bit AND instruction is stored in the internal buffer of decoder unit. Next cycle is execution cycle for *MUL* instruction in which multiplication operation is carried out as per the received function code *0D*. Note that, multiplication operation is performed by repetitive addition algorithm. In this algorithm, content of operand1 is added with itself until content of operand2 (multiplicand) reaches to zero. As observed from the simulation results, execution of 16-bit *MUL* operation is completed within a single pipe line cycle only. As shown in Fig. 5.14, after execution of an instruction the least significant 32-bit of the result is stored. After computation, as per result *FFFFFF01H,* the negative flag is updated. At the negative edge of same clock cycle, content of both *PSR* register (*80000000H*) and destination register *R5* are updated (*FFFFFF01H*) as per control signals generated from the control unit.

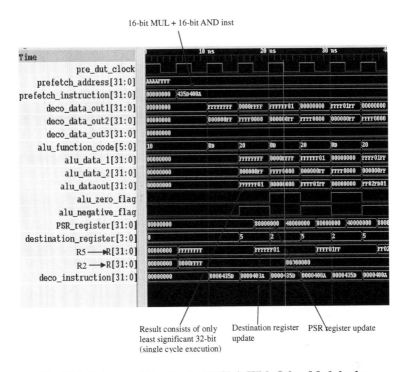

16-bit MUL + 16-bit AND inst

Result consists of only least significant 32-bit (single cycle execution)

Destination register update

PSR register update

Fig 5.14 Integrated Result of *ALU* Unit With Other Modules by Considering Execution of 16-bit MUL and 16-bit AND Instructions

Subsequent clock cycle is execution cycle for 16-bit *AND* instruction. As per received function code *20*, *ALU* performs the logical ANDing operation between two received operands and computes the result. As the result after computation is *00*, zero flag is set and PSR register is updated with the value *80000000H* at negative edge of the same clock cycle. The destination register *R2* is also updated with value *00* at the negative edge of clock cycle by control unit.

Fig.5.15 illustrates simulation results of execution of conditional branch instruction functionality. First, the value *000FFFFFH*, *FFFF0000H* and

109

FFFF0001H are stored into registers *R0*, *R1* and *R15* respectively using test vectors *reg_no*, *tb_write* and *tb_data_in*. Then, prefetch address *AAAB0000H* with op-code of 16-bit AND instruction followed by opcode of 16-bit conditional branch instruction *400AD0FFH* and op-code of 16-bit X-OR instruction followed by 16-bit *OR* instruction *40714033H* are transferred to pre-fetcher using test vector *tb_pre_prefetch_address* and *tb_pre_prefetch_instruction*. As per the pre-fetch address *AAAB0000H*, opcode for the 16-bit *AND* instruction followed by the 16-bit conditional branch instruction are received by the pre-fetch stage in the second cycle. In the next cycle, decoding operation is carried out for 16-bit AND instruction and also necessary function code "20" is determined for AND operation. In the same clock cycle the op-code for 16-bit conditional branch instruction is stored in the internal buffer of decoder unit (D0FF). Subsequent cycle is execution cycle of 16-bit AND instruction. So, *ALU* performs the logical ANDing between the two received operands and computes the desired result for received function code *20*. As the result after the computation is zero, the zero flag is set by the instruction; and subsequently PSR register (*40000000H*) and destination register *R2* are updated (00000000H) at the negative edge of same clock cycle. In the same cycle, op-code is decoded for 16-bit conditional branch instruction and branch target address *FFFF0104H* is computed in order to fetch the branch target address instruction (*16-bit OR* instruction). This feature in *ARM Cortex - M3* processor is referred as a branch forwarding. During decoding operation, the condition code *0* of branch instruction is determined to indicate the condition required to be checked for branch execution.

The next cycle is execution cycle of 16-bit conditional branch instruction. During execution, *ALU* first checks condition of zero flag for branch execution. As the previous *16-bit AND* instruction has already set the zero

110

flag, the condition (decode stage) is found to be true and branch execution takes place. As the branch is executed, the content of PC and offset value are added; and control signal *branch_execution* is set high.

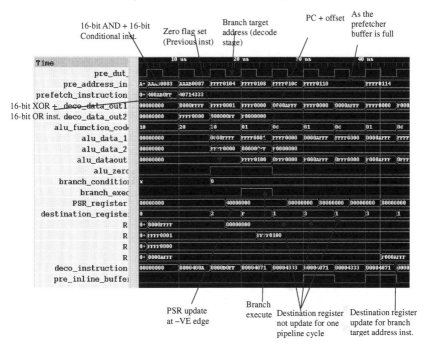

Fig 5.15 Integrated Result of ALU Unit With Conditional Branch Instruction Executed

branch execution signal is used to instruct the decoder and pre-fetcher unit to suspend the normal pre-fetching operation and the pipeline reloading operation is required to continue for one complete pipeline cycle [17]. Note that pipeline reloading means the next sequential instructions are executed but their destination registers are not updated. As the pipeline reloading operation requires one pipeline cycle, three subsequent instructions (16-bit X-OR, 16-bit OR and 16-bit X-OR) are executed but their corresponding

111

destination registers (*R1, R3 and R1*) are not updated. Destination register is updated only for the fourth instruction (16-bit OR) that is a branch target address instruction. For this instruction, destination register R3 is updated with the value *F000AFFFH* at the negative edge of the clock cycle. *PSR* register is updated for the fourth instruction at the negative edge of the clock cycle. Simulation result shown below in Fig 5.16 illustrates functionality of 16-bit conditional branch instruction not executed. As discussed above, the branch target address is computed in the decode stage itself. As 16-bit AND instruction has not updated the zero flag in the previous clock cycle, branch execution does not take place in execution stage and control signal *branch_execution* is reset to low. Consequently, current value of Program Counter (*PC*) is preserved (*FFFF0001H*). Since branching operation has not taken place, sequential execution is continued from the address *AAAB000BH* onwards. As shown in the simulation results, the destination register *R1* is updated for the next sequential *16-bit XOR* instruction with the value *0FFF0000H*.

As shown in Fig.5.17, execution of 16-bit unconditional branch instruction. Prefetch address *AAAB0000H* with op-code of 16-bit *AND* instruction followed by op-code of 16-bit unconditional branch instruction *400AE7FFH* and op-code of 16-bit *X-OR* instruction followed by 16-bit *OR* instruction *40714033H* are transferred to pre-fetcher using test vectors *tb_pre_prefetch_address* and *tb_pre_prefetch_instruction*. Like conditional branch instruction, the branch target address (*FFF0104H*) for branch target address instruction is computed in decodes stage to pre-fetch the branch target address instruction. In the next clock cycle, unconditional branch execution is performed by adding value of *PC* with the offset value. In the same clock cycle, signal *brach_execution* is set to be high. b*ranch_execution* signal is used to instruct the decoder and pre-fetcher unit to suspend the normal pre-fetching operation and pipeline reloading operation takes place for one complete pipeline cycle [17].

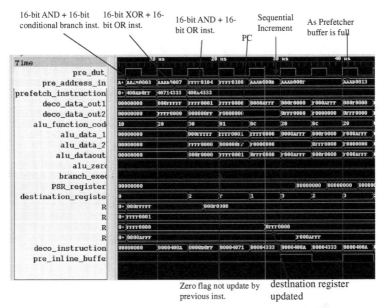

Fig 5.16   Conditional Branch Instruction Not Executed

As discussed while pipeline reloading, the next sequential instructions are executed but their destination registers are not updated. Since pipeline reloading operation requires one pipeline cycle, three subsequent instructions are executed without updating corresponding destination registers (*R1, R3 and R2*). The destination register is updated only for the fourth instruction (*16-bit OR*), which is a branch target address instruction. For this instruction, destination register *R3* is updated with value *F000AFFFH* at the negative edge of the clock cycle. *PSR* register is updated (*80000000H*) for the fourth instruction at negative edge of the clock cycle.

113

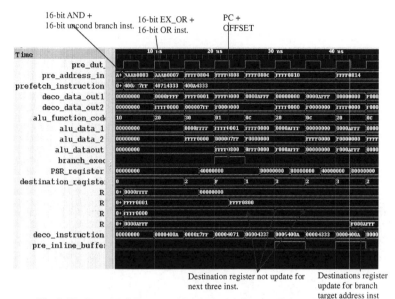

16-bit AND +
16-bit uncond branch inst.

16-bit EX_OR +
16-bit OR inst.

PC +
OFFSET

Destination register not update for
next three inst.

Destinations register
update for branch
target address inst

**Fig 5.17 Result of Integrated ALU with Unconditional Branch
Instruction**

LOAD/STORE with multiple registers

Fig.5.18 shows the functionality of Load/Store with multiple registers. Initially write operation is initiated using test vector *tb_HWRITE*. High status of *tb_HWRITE* indicates write bus transfer. Then, initial address *000000A0H* and opcode *CE07CC07H* of back to back *LDM* instruction is transferred to memory using test vector *tb_HADDR* and *tb_HWDATA*. As a result, op-code is stored at memory location *0000000A0H*. Data *00000004H* is stored into registers *R6* using test vectors *reg_no*, *tb_write* and *tb_data_in* to determine the base address. In order to initiate read operation, the initial Pre-fetch address *000000A0H* is given to Master using test vector *tb_pre_prefetch_address*.

Consequently, Master initiates read operation and op-code of *LDM*

114

instruction is received by pre-fetcher unit and in same cycle, data is forwarded to fetch stage. During decode clock cycle, decoding operation takes place for the first 16-bit LDM instruction (*CE07*) and second 16-bit LDM instruction (*CC07H*) is required to be stored into internal buffer of decoder unit. In the next cycle, the base address (*00000004H*) is computed by Load/Store address unit as per value of *operand1*. Base address is the address, transferred to memory in order to fetch the data available at that address (*00000004H*). According to the nature of AHB bus, data is received by LSA execution unit in the cycle after the next cycle. At the negative edge of clock cycle, register *R0* is updated with the value *88888888H* and at the subsequent negative edge of the clock cycles, registers *R1* and *R2* are updated with the value 99999999H and *11111111H*.

Advanced High Performance Bus Unit (AHB)

AHB bus is an external bus used to transfer code and data from memory. *AHB* Bus protocol consists of one Master and multiple Slave devices. In this project, pre-fetcher serves as a Master and Memory serves as a Slave device.

Fig.5.19 illustrates the simulation results of functionality of write and read bus transfer. *HWRITE, HADDR* and *HWDATA* signals are driven form test cases in order to validate functionality of write/read bus transfer. Write operation is initiated using test vector *tb_HWRITE*. High status of *tb_HWRITE* indicates write bus transfer. Then, initial address 00000000H is transferred to memory using test vector *tb_HADDR* and address is incremented with the factor of *+4* (*00000004H* and *00000008H*). Once the initial address is transferred to memory, op-code to be stored into memory is transferred using test vectors *tb_HWDATA*.

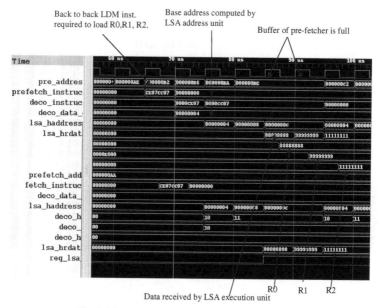

Fig 5.18   Load/Store With Multiple Registers

As a result, the first op-code *EA110242H* is stored at memory location *00000000H*. Remaining op-codes *400A400AH* and *40714071H* are stored from memory locations *00000004H* and *00000008H* respectively. In order to initiate read operation, initial Pre-fetch address *00* is given to Master using test vector *tb_pre_prefetch_address*. At positive edge of the fifth clock cycle, the Master has initiated read transfer by sending starting address *00000000H* on *HADDR* bus and generating various control signals like *HTRANS* with the value *10* that indicates *NONSEQ* address and *HSIZE* control signal with the value *10* that indicates word size transfer. Further, at the positive edge of the next clock cycle, the slave first samples the address (*00000000H*) and transfers data (*EA110242H*) on data bus in the same cycle. Note that, after transferring data on data bus, the slave generates high *HREDAY* control signal, which indicates that data transfer has completed on

116

data bus. The slave also generates additional control signal *HRESP* with the value *10* to indicate that transfer is progressing normally. In the same clock cycle, master changes status of control signal *HTRANS* with the value *11* that indicates SEQ address. On the subsequent clock cycles, the slave transfers opcodes (*400A400AH*) and (*40714071H*) respectively on data bus. Data available on the data bus is received by pre-fetcher unit in the next clock cycle.

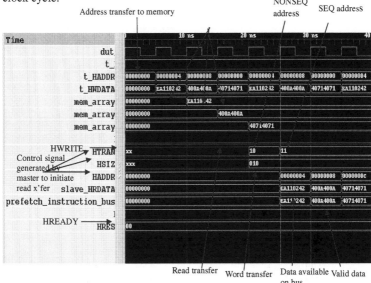

Fig 5.19 Write and Read Bus Transfers

Integration of core with AHB

As shown in Fig.5.20, the simulations results depict the integration of core with the AHB bus protocol. Initially, data *FFFF000FH* and *0000FFFFH* are stored into registers *R1* and *R2* respectively using test vectors *reg_no*, *tb_write* and *tb_data_in*. Write bus transfer is initiated using test vector *tb_HWRITE*. As shown in the figure, high status of *tb_HWRITE* indicates write bus transfer. Initial address *0x08H* and opcode of *32-bit AND* instruction *EA110242H* are sent to memory using test vectors *tb_HADDR*

117

and *tb_HWDATA* in order to store op-code at a memory location *0x08*.

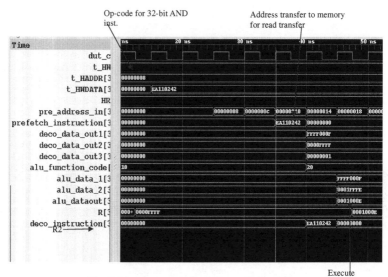

Fig 5.20 Integration of Core With AHB

To initiate read bus transfer, pre-fetch address *0x08* is sent to Pre-fetcher using test vectors *tb_pre_prefetch_address*. At the positive edge of fourth clock cycle, Master initiates read transfer by sending starting address *00000008H*. As per the nature of AHB bus, op-code is received by Pre-fetcher unit in the cycle following the next clock cycle. In the subsequent cycle, op-code decoding takes place and also the function code *20H* is determined to carry out further operations. The next cycle is execution cycle of pipeline in which execution of *AND* operation takes place and destination register *R2* is updated with the value *0001000EH* at the negative edge of the same clock cycle.

118

Cortex – M3 processor consists of a number of system exceptions and external exceptions. Vector address of corresponding exception is available in a vector table. Vector table with the address *0x00* contains initial value of SP_Main and the address *0x04* contains vector address of Reset ISR. The vector table in the memory address 0x00 contains value *0000FFFFH* and the memory location *0x04* contains *0x000000AAH,* which is a vector address of Reset *ISR.*

Fig. 5.21 illustrates Exception modeling for Reset exception. Write bus transfer is initiated using test vector *tb_HWRITE.* As shown in the figure, *tb_HWRITE* indicates write bus transfer. Then, address *0x00* with data *0000FFFFH* are sent to memory using test vectors *tb_HADDR* and *tb_HWDATA* in order to store initial value of SP_Main and vector address of Reset ISR. Reset exception is asserted using test vector *tb_reset* and then it is disserted in the next clock cycle. As shown in the figure, in the first clock cycle, status of *reset_exe* signal is high that indicates Reset exception has occurred. Consequently, the module transfers address *0x00* in cycle after the next clock cycle on the address bus in order to fetch the initial value of SP_Main from the vector table. Then, the address is incremented by *+4* in order to fetch the vector address of Reset ISR from the vector table. Once the address *0x00* is transferred to memory, the data available on that address is available to the bus in cycle following the next clock cycle as per the nature of AHB bus. At the negative edge of same clock cycle, register *R13* (SP_Main) is updated with the value *(0000FFFFH),* which defines top of stack for the stack pointer *(SP).* At the negative edge of the subsequent cycle, Program Counter *(PC)* register is updated with the value *0000000AAH,* which is vector address of Reset ISR. As observed from the simulation results, address on *pre_address_in* bus changes from *0000000CH*

119

to *000000AAH* and indicates that the next instruction to execute is the first instruction of Reset *ISR*.

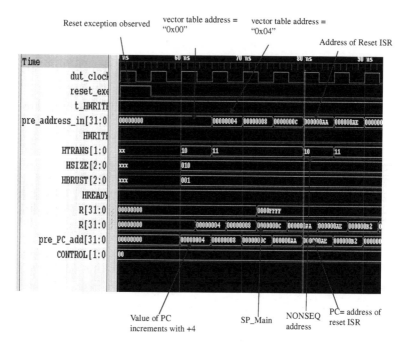

Fig. 5.21 Results of Reset Exception

5.3   SUMMARY

The chapter begins with the simulation environment adopted for this project. Major focus of this chapter is on discussing the simulation results, which describe behavior of various stages of pipeline architecture of the core. Simulation results also trace execution timing of various Thumb and Thumb-2 instructions. The chapter discusses verification methodology used during project work by considering different test vectors.

120

CHAPTER 6
CONCLUSION AND FUTURE WORK

6.1 CONCLUSION

ARM Cortex – M3 processor is latest recent development in the series to support *VLSI/ASIC* based embedded system development in industries. The model of *ARM Cortex – M3* processor generated by this project is expected to serve as a master device in order to verify functionalities and behavior of a number of slave devices.

The activity has been carried out using *SystemC* as a modeling tool as summarized below.

- Architecture blocks of different stages/units like Instruction Fetch, Instruction Decode, Shifter, *ALU*, Load-Store, *AHB* and memory have been modeled by considering various control signals in accordance with the design specifications.
- Functionality of each module/unit has been successfully simulated.
- Results are viewed on waveform viewer as well as in executable file for debugging the errors.
- Functionalities are verified by passing different test conditions through test bench.
- It is observed that the model is able to perform all operations in the number of clock cycles as indicated in manufactures specifications.
- All modules are integrated and their functionalities are verified again with reference to various timing relationships after verification.

Majority of Thumb and Thumb-2 instructions have been verified as the co-processor is not included in the model.

FUTURE SCOPE

There are more than *200* exceptions in the design of *ARM Cortex – M3* processor. Scope of this project has been restricted to generate model of the processor without exceptions. However, we have implemented Reset exception in the model. In future, complete exception modeling may be attempted.

REFRENCES

[1] William Stallings, "Reduced Instruction Set Computer Architecture," in Processding of IEEE Journal, Vol. 76, No.1, pp. 38-55, Jan. 1988.

[2] Paul Chaw, "RISC (Reduced Instruction Set Computer)," in journal of IEEE POTENTIAL, pp. 28-31, OCT- 1991, 00278-6648/91/0010-0028.

[3] D. Patterson and D. Ditrel, "The Case for the RISC," in Computer architecture News, Vol. 8, No. 6, pp. 25-33, Oct. 1980.

[4] John Hennessy, Norman Jouppi, Forest Baskett, Thomas Gross, & John Gill. "Hardware Software Tradeoffs for Increased Performance," in journal of Symposium on Architectural Support for Programming Languages and Operating Systems, pp. 2-1 1, Mar. 1982.

[5] Sivarama P. Dandamudi, "Guide to RISC Processors for Programmers and Engineers; Chapter 3; "RISC Principles", Prentice-Hall, Mar. 2005, ISBN 978-0-387-21017-9."

[6] D. Bhandarkar and D. W. Clark, "Performance from Architecture: comparing a RISC and a CISC with similar Hardware Organization," Proc. 4th International Conference on Architectural Support for Programming Languages and Operating Systems, 1991, pp. 310-319.

[7] Samuel O. Aletan, "An Overview of RISC Architecture," Proc. Symp. Applied Computing, 1992, pp. 11-12.

[8] Steve furber, "ARM System- on-chip architecture" second edition, Pearson Edition, Mar.2000,

[9] Andrew N.SLOSS, Dominic. SYMES, Chris WRIGHT, "ARM System Developer's Guide Designing and Optimizing System Software", Morgan Kauffman Publishers, Mar-2004, ISBN: 1-55860-874-5.

[10] Patterson, David A: John L. Hennessy. Computer Organization and design: "The Hardware/Software Interface" Morgan Kaufmann Publishers. ISBN 1-55860-604-1.

[11] Mamun Bin Ibne Reaz,, Md. Shabiul Islam, and S. Sulaiman, "A Single Clock Cycle MIPS RISC Processor Design using VHDL" in Proc. ICSE 2002, Malasiya. pp, 199-203.

[12] Various SPARC v7 implementations were produced by Fujitsu, LSI Logic, Weitek, Texas Instruments and Dypress.

[13] Liam Goudge and Simon Segars, "Thumb: Reducing the Cost of 32-bit RISC Performance in Portable and Consumer Applications", in IEEE Proceedings of COMPCON '96, pp. 176-181.

[14] Joseph Yiu, "The Definitive Guide to the ARM Cortex– M3", Newnes publishers, ISBN: 978-0-7506-8534-4.

[15] Improving ARM code Density and performance. New Thumb extension to the ARM architecture by Richard Phelan June 2003. Thumb-2 white paper.

[16] Shyam Sadasivan, "An Introduction to ARM Cortex-M3 Processor", in Proc. of International conference, October 2006,

[17] Cortex™- M3 Revision: r1p1 Technical Reference Manual.

[18] Samir Palnitkar, "Verilog HDL, A Guide to Digital Design and Synthesis", Second Edition IEEE 1364-3001, Compliant by Pearson edition, 2001.

[19] A SystemC ™ Primer By J. BHASKER Cadence Design Systems. Star Galaxy Publishing.

[20] AMBA™ Specification (Rev 2.0)

[21] Willam Stallings, "Computer Organization and Architecture", Pearson edition, sixth edition (Design and Performance), 1996.

[22] ARM v7-M Architecture Application Level Reference Manual. ARM DDI 0405C.

[23] "SystemC – "A modeling platform supporting multiple design abstractions" by Preeti Ranjan Panda Synopsys Inc. 700 E. Middlefield Rd. Mountain View, CA 94043, USA.

[24] IEEE Standard SystemC Language Reference Manual, IEEE Standard 1666-2005.

[25] John Goodacre, Andrew N.Sloss, "Parallelism and the ARM instruction Set Architecture" in IEEE Computer Society -2005, pp. 42-50.

[26] ARM7TDMI (Rev 3) Technical Reference Manual.

[27] "Sample SystemC Design" ALDEC Whitepaper Excerpt by Jaroslaw Kaczynski Ver.1.2-2005-08-17.

APPENDIX -A

SYSTEMC

A.1 SYSTEMC AS A MODELING LANGUAGE

Now day's designs are getting bigger and bigger in size, faster in speed and larger in complexity. This requires describing the design at the higher level of abstraction so as to enable:

- Faster simulation.

- Architectural exploration.

- Hardware/Software co simulation.

Expressing designs at the system level becomes also important to manage complexity of design in order to perform all design optimization and exploration at system level. When design is expressed at system level it is very easier to explore various algorithms and alternate architectures as and when required as compared to exploring at register transfer level or gate level. At chip level size of design is larger so it is quite difficult to explore various design changes and time consuming too. At system level size of design is manageable and different design changes and different architectures changes can be made easily done. The unique thing that SystemC is best suited for modeling is that the same language is used for System level design, describing hardware architectures, describing software algorithms, verification and IP exchange [19]. Means we can write design in one language, verify the design using same language and further refine it all the way to the implementation level. We can describe overall system using *SystemC*. As *SystemC* is based on *C++* hence it is powerful enough to

126

describe all kinds of verification environment from the signal level to transaction level. It also allows for test bench generation and testing so it can serve as a verification language as well. *SystemC* also provides a single unique language for IP creation either at the register transfers level or at the system level, including the test environment. In addition, the *SystemC* allows for a full standard simulation environment that the user can use to verify the IP. *SystemC* has direct support for modeling at the high level of abstraction. Then it will redefine at behavioral level or register transfer level using the same language.

SystemC supports design abstraction at the *RTL*, behavioral and system levels. Consisting of a class library and a simulation kernel, the language is an attempt at standardization of a *C/C++* design methodology. It is supported by the Open SystemC Initiative (*OSCI*), a consortium of a wide range of system houses, semiconductor companies, IP providers, embedded software developers, and design automation tool vendors [27]. Apart from the modeling benefits available in *C++* such as data abstraction, modularity, and object orientation. The advantages of *SystemC* include the establishment of a common design environment consisting of *C++* libraries, models and tools, thereby setting up a foundation for hardware-software co-design; the ability to exchange IP easily and efficiently. It also has ability to reuse test benches across different levels of modeling abstraction [23].

A.2 FEATURES OF SYSTEMC

SystemC consists of following important features which are extremely useful during modeling of architecture:

- *SystemC* is based on the *C++* programming language. *SystemC*

127

extends the capabilities of C++ by enabling modeling of hardware descriptions. It adds such important concepts to C++ as concurrency means multiple processes execute concurrently, timed events and data types.

- *SystemC* adds a class library to C++ to extend the capabilities of C++. Class library is not a modification of C++ but it contains library of functions for ease description, data types and other language constructs which are legal C+ code. Class library provides new mechanism to model any architecture with concurrency, timing constraints and reactive behavior. Class library provides various constructs that determines the concepts such as modules, ports, channels, and signals. The library also provides certain built in functions which are extremely useful during modeling of any architecture. As *SystemC* defines a new C++ class library without adding any new syntax to C++ programming language. It is simply C++ based language, these classes enables the user to define modules, processes, and channels that are communicated through ports and signals which can handle a wide range of data types ranging from bit vectors to logic vectors and also user defined data types such as enumeration types and structures types.

- *SystemC* also provides a simulation kernel that allows you to simulate the executable specification of design or code that written in *SystemC* by writing simulation syntax such as sc_start and sc_stop. *SystemC* also provides a single language to define hardware and software components. It also provides single language to facilitate hardware and software cosimulation. It also provides a single language for step by step mapping from system level design to register transfer level design for synthesis purpose. As *SystemC* is based on standard C++ programming language, we can use the C++

environment tool to crate a system level design, compile, and simulate the design as quickly as possible to validate and optimize the design.

- *SystemC* also contains rich data types ranging from bit vectors to logic vectors which supporting *X* values and *Z* values for equating or assigning purpose. Also it supports user defined data types such as enumerated data types and structure data types.

- *SystemC* also provides tracing functions for variables and signals, using which we can debug our design easily.

- Further more *SystemC* is an open source and it is a freely available language [19].

A.3   THE SYSTEMC DESIGN FLOW

A.3.1   Simulation with SystemC

Fig. A.1 below illustrates a typical simulation methodology in the SystemC environment. The designer writes the *SystemC* models at the system level, behavioral level, or *RTL* level using *C/C++* augmented by the *SystemC* class library. The class library serves two important purposes. First, it provides the implementation of many types of objects that are hardware-specific, such as concurrent and hierarchical modules, ports, and clocks. Second, it contains a kernel for scheduling the processes.

The *SystemC* code of user can now be compiled and linked together with the class library with any standard *C++* compiler (such as GNU's gcc), and the resulting executable serves as the simulator of the user's design.

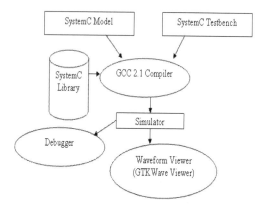

Fig. A.1 Simulation with SystemC

The test bench for verifying the correctness of the design is also written in SystemC and compiled along with the design. The executable can be debugged in any familiar *C++* debugging environment (such as GNU's gdb). Additionally, trace files can also be generated to view the history of selected signals using a standard waveform display tools. The program development infrastructure already in place for *C/C++* can be directly utilized for the *SystemC* verification and debugging tasks. For hardware designers used to viewing simulation data in the form of waveform displays, the trace file generation facility provides a familiar interface. Conceptually, the most powerful feature is that the hardware, software, and test bench parts of the design can be simulated in one simple and unified simulation environment without the need for clumsy co-simulations of disparate modeling paradigms [23].

A.3.2 Implementation and Synthesis.

The important feature of the *SystemC* implementation flow is that the specification is in a common language for both hardware and software parts so implementation of different parts of design in hardware or software can be explored in seamless fashion, eliminating the need of rewriting each module in both *C* and *HDL* languages. The design entry could be at any level of abstraction: system level, behavioral level, or *RTL* level, the transition from a higher level to a lower level of abstraction could be achieved either through automatic synthesis and compilation tools such as hardware/software partitioning and co-synthesis tools for determining which portion of the design is synthesized into gates and which portion is compiled into embedded software; and behavioral synthesis tools, or through a manual refinement process. Finally, the *RTL*-level design, whether generated by hand or by previous synthesis steps, is the input to *RTL*- and logic-synthesis tools familiar to hardware designers; the output is a gate-level net list. The specification language remains the same across all levels of synthesis, and the changes in abstraction level involve a refinement into greater detail within the same language and design environment. This allows the same test bench to be used to verify the design at multiple levels, if carefully designed, resulting in a design environment that is very tightly integrated. Of course, since C++ has many constructs that are unrelated to hardware, appropriate subsets will have to use for synthesis.

A.4 HARDWARE MODELING

In order for language to be acceptable as a designing language at multiple abstraction level, it is important that its constructs are analogues with the current Hardware Description Language (*HDL*) like Verilog and *VHDL*. SystemC provides mechanism to model the typical hardware functionality by means of constructs that are defined by class library which are analogues to current *HDLs*.

A.4.1 Structure and Hierarchy

Modules

Structural decomposition is one of the fundamental hardware modeling concepts because it helps partition a complex design into smaller entities. A module is a basic unit for describing structure in and that are the major building blocks in *SystemC*. A *SystemC* description consists of a set of connected modules, each encapsulating some behavior or functionality. Modules can be hierarchical, containing instances of other modules. The nesting of hierarchy can be arbitrarily deep, which is an important requirement for structural design representation. A module may consist of any number of processes, module is defined by construct *SC_MODULE* in *SystemC*.

Signals and Ports

The simplest means of connecting together different SystemC modules are by using ports and signals. Actually, the interface of modules to the external world can be much more general but the interface at the lowest level matches with the typical facilities available with the current *HDLs*. A port has direction like input, output or inout (bidirectional). Following Fig.A.2 shows a simple structural design consisting of a one module C with instantiations of two modules A and B within it which is named A1 and B1 respectively with the following specifications:

132

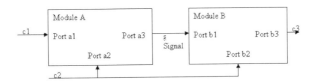

Fig. A.2 Simple structural design contains Modules, ports, signals.

As shown in above Fig. Module A consist of, two input port a1 and a2, one output port a3. Module B consists of two input ports b1 and b2, one output port b3. Module C consists of two input ports c1 and c2, one output port c3. Note that signal s is used to connect port of Module A and B. The SystemC description for this structure is as follows:

SC_MODULE (A) {// Module declaration.

sc_in<bool> a1; // Port declarations.

sc_in<bool> a2; // Port declarations.

sc_out<bool> a3;// Port declarations.

}; // Module end.

SC_MODULE (B) {// Module declaration.

sc_in<bool> b1;

sc_in<bool> b2;

sc_out<bool> b3;

};

SC_MODULE (C) {

sc_in<bool> c1;

sc_in<bool> c2;

sc_out<bool> c3;

A *A1;// instances of module A.

B *B1;// instances of module B.

sc_signal<bool> s; // signal declaration

```
SC_CTOR (C) {// constructor
A1 = new A ("A1"); // Module instantiation.
(*A1) (c1, c2, s); // Port mapping.
B1 = new B ("B1"); // Module instantiation.
(*B1) (s, c2, c3); // port mapping.
}
}; //end of structure.
```

As illustrate in above description, Module is describe with keyword
SC_MODULE and port are describe with sc_in, sc_out and sc_inout
keywords. The data types used over here is defined by <bool> indicating
that type is Boolean or single bit. The whole structure hierarchy is
specified in under constructor that defines with the key word SC_CTOR.
Modules A1 and B1 are instantiated with the help of pointer with construct
new. The port mapping is done by connecting port c1, c2 and signal s to
three ports of Module A1. Note that signal s is used as a wire for connecting
the output port a3 of A1 to input port b1 of B1.

A.4.2 Functionality and Concurrency

The process is used to describe functionality of a system in SystemC. As
similar to VHDL processes, the SystemC processes are used to represent
concurrent behavior means multiple processes within a module represent
hardware or software blocks executing in parallel. Processes contains
sensitivity list that are list of signals that trigger the execution of the
process. In SystemC, there are two important types of processes, Method
and Thread, process are registered in a constructor implying it as a
recognized process. A module can have multiple SC_METHOD processes.

Each process can either model combinational logic or can model synchronous logic. Communication between processes occur using signals.

Methods

A method process behaves like a function call and can be used to model simple combinational behavior. It does not have its own thread of execution, and hence, cannot be suspended. This characteristic allows for high simulation efficiency.

Threads

A thread process can be used to model sequential behavior. It is associated with its own thread of execution, and can be suspended and re-activated. A simple example involving method and thread processes is shown below. Functions m and n are registered as a method process and a thread process respectively. The sensitivity list is specified using the sensitive keyword and the << operator.

```
SC_MODULE (X) {
sc_in<bool> a, b;
void m(); // Function definition omitted
void n();


SC_CTOR (X) {
SC_METHOD (m); sensitive << a;
SC_THREAD (n); sensitive << a << b;
}
};
```

A.4.3 Times and Clocks

Since in hardware modeling the concept of time and clock are very important, SystemC provides constructs to specify them using followings:
sc_clock clk ("clk", 5, SC_NS);

Above syntax specifies the clock signal with time period of 5 nano- second. sc_clock is a keyword used to represent the clock signals. The process can be synchronized with clock signal using defining clock signal in sensitive list. The sensitive, sensitive_pos, sensitive_neg are used to specify synchronization of a process to a clock using the following syntax:
SC_THREAD (x);
sensitive_pos << clk;
It ensures that process x is activated on the positive edge of clock signal clk.

A.4.4 TESTBENCH

SystemC is a language not only describing the hardware but also for building strong verification test environment. It is C++, so all features of C++ are used to write any sort of complex verification model. A test bench is a model used to verify the correctness a design under test; a test bench is also using the same language, that is, in SystemC.

A test bench written for major three purposes:

- To generate stimulus for simulation.

- To apply stimulus to design under test and determine the response.

- To compare the output response with excepted results.

In SystemC, a test bench is specified with an SC THREAD, just like any other process, and is easily integrated into the overall design. Sophisticated test benches can be built using all the constructs available in C++. Since the test bench does not need to be synthesized, there is no need to conform to any synthesizable subset of C++ while writing them.

A.5 DATA TYPES

SystemC provides a rich set of data types which can used to model hardware specific concepts in addition to standard data types of C++ like bool, int, char etc. Some of the useful data types are mentioned below:

Bit and Bit Vector:
The sc bit and sc bv types can be used to model bits and bit vectors for which only two states, '0' and '1' are valid, and on which logical operations such as logical AND, logical OR, etc are performed. Useful operations for these types include the reduction like and reduce, or reduce, and xor reduce and part-select (range). For example,

sc_bv<150> x, y; // Indicates   150-bit vectors.
a = a | b; // Indicates logical OR operation.
sc_logic r = m.and_reduce(); // Indicates AND reduction operations.
sc_bv<50> z = y.range (49, 0); // Indicate range or part selection operation.

4-state Logic or arbitrary logic type:
In addition to the standard bit values '0' and '1', it is useful to provide a mechanism to indicate that the value of a bit is unknown. This helps identify initialization or conflict (multiple driver) problems during simulations. Further, there is the need to specify the high impedance state on signals. With this in mind, SystemC provides sc logic, a four state logic data type, the states being '0' (low or false), '1' (high or true), 'X' (unknown), and 'Z' (high impedance). A logic vector data type, sc_lv, is used to specify data

items more than one bit wide that need to be modeled with 4-state logic, e.g., a tri- statable data bus. A bus can be tristated as follows:

sc_lv<8> data; // 8 bits wide

data = "ZZZZZZZZ"; // set to high impedance

SystemC also provides data types to represent resolved logic signals which are useful in modeling wires and buses with multiple drivers.

Fixed and Arbitrary Precision:

The integer data type's int and unsigned used in C++, have an implementation dependent bit width. However, the designer may wish to fix the precision of a data item if the range of values it takes is known in advance. SystemC provides two data type families for achieving this: fixed precision and arbitrary precision. The fixed precision types are widely used in modeling like sc int and sc uint that is up to 64 bits wide. These data types are implemented with a 64 bit integer. Generally the usual operations associated with C++ integers can be applied to the fixed precision types, with these data types one of the useful operation being used that is bit-select operation. For example:

sc_int<48> x; // Indicates signed, 48 bits wide data.

sc_uint<40> y; // Indicates unsigned, 40 bits wide data.

sc_logic p = x[3]; // bit select

The regular arithmetic operations can be performed on these types.

There may be cases where a bit-width larger than 64 bits is needed to model some data, for example, a wide data bus. In such cases, the arbitrary precision types, sc bigint and sc biguint provided by SystemC can be utilized. For example,

sc_bigint<128> x; // signed, 128 bit

Fixed Point Representation:

The sc fixed and sc unfixed data types, which are used to represent such fixed point numbers in SystemC, are accompanied by the standard characteristics of fixed point arithmetic, such as quantization mode, overflow mode, and saturation bits.

sc_fixed<10, 7, SC_RND, SC_WRAP, 3> a;

The above construct used to define a variable a with total word length of 10 bits, integer word length of 5 bits, quantization mode which is round to plus infinity , overflow mode is wrap around and 2 saturation bits. The fixed point data types is an important modeling features of SystemC that is not found in any other HDLs.

The choice of data type among the different data types has a grater significant importance in modeling, also it has an impact on the simulation speed, and care must be taken to choose the correct data types during modeling. For example, sc_lv should be used only in specific instances where either high impedance behavior is involved, or reset behavior in simulation is important; otherwise, the faster sc_ bv type should be used. Similarly, if extensive arithmetic is performed, the sc_int and sc_bigint types should be preferred to sc_bv to prevent unnecessary data type conversions. The fixed precision types sc_int should be used wherever possible instead of sc_bigint for simulation efficiency. Finally, the int native C++ types are the most efficient.

A.6 SYSTEM LEVEL MODELING

The most advantageous use of SystemC is, it contains all the required hardware modeling features with that it provides powerful modeling constructs for system level design. This ensures that the transition to a

SystemC-based methodology entails no compromise in terms of expressive power at the lower levels of abstraction, and yet provides a useful framework for modeling at the higher levels. The system level modeling features introduced in SystemC 2.0 [23] mainly include the support of a much more general and abstract means of communication between processes and a more general mechanism for event synchronization.

A.6.1 Events and Sensitivity

SystemC introduces a general mechanism for specification and notification events. The sc_event can be used to declare events that can be defined by using notify keyword and be synchronized with in wait statements, as shown below.

sc_event a1, a2; // declare events.

sc_time t (5, SC_NS);

e1.notify (t); notify event after 5 ns.

wait (a1); suspend execution until a1 events occur.

A.6.2 Primitive and Hierarchical Channels

Channels in SystemC can be either primitive or hierarchical. Primitive channels are relatively simple; SystemC 2.0 provides a set of primitive channels which have wide applicability, such as sc_ signal , sc_mutex used to model mutual exclusion and sc FIFO used for modeling queues. Hierarchical channels can exhibit structure; they are modules themselves, which in turn, can contain processes, and other channels and modules.

A.7 SIMULATION KERNEL

The simulation kernel for SystemC follows the evaluate-update paradigm that is common in HDLs. The concept of delta cycles, where multiple evaluate-update phases can occur at the same simulation time, is supported [24]. A simplified version of the simulation algorithm is as follows:

- Initialization: Execute all processes to initialize the system.
- Evaluate: Execute a process that is ready to run. Iterate until all ready processes are executed. Events occurring during the execution could add new processes to the ready list.
- Update: Execute any update calls made during step 2.
- If delayed notifications are pending, determine list of ready processes and proceed to evaluate phase (step 2).
- Advance the simulation time to the earliest pending timed notification. If no such event exists, simulation is finished, else determine ready processes and proceed to step 2.

Made in the USA
Lexington, KY
31 January 2017